COUNTER-INSURGENCY CAMPAIGNING

COUNTER-INSURGENCY CAMPAIGNING

JULIAN PAGET

with a Foreword by

LIEUTENANT GENERAL D. B. LANG C.B., D.S.O., M.C.,
Commander-in-Chief, Scottish Command
former Director of Army Training

FABER AND FABER LIMITED
24 Russell Square London

First published in mcmlxvii
by Faber and Faber Limited
24 Russell Square London WC1
Printed in Great Britain by
Latimer Trend & Co Ltd Plymouth

CONTENTS

FOREWORD

by Lieutenant General D. B. Lang, C.B., D.S.O., M.C.
former Director of Military Training

Since the end of the Second World War, we have been training and preparing our fighting forces for a third conflict, possibly even more disastrous than any yet, because it might include nuclear weapons. But at the same time, British and many other troops have been engaged in active operations in many parts of the world. From British Guiana to Vietnam, security forces have been waging a relentless war against insurgent elements; and there seems to be no end in sight. In fact it has been suggested that this *is* the Third World War, and that this will be the pattern of international conflict for at least a generation to come. Certainly the pattern of events over the last twenty years suggests that we have not seen the end of 'small wars.'

Algeria, Kenya, Cyprus and Malaya, to mention but a few, have all been the scene of bitter and prolonged insurgency. Much has been written about them, particularly by journalists, who have produced vivid pictures of the scene and the personalities. But very few of these writers have had the professional knowledge to search further below the surface to analyse the mechanics of this modern type of warfare.

It is to this problem that Colonel Julian Paget has turned his attention. As a professional soldier who has been concerned at one level or another in several counter-insurgency campaigns, he has made a study from a military point of view of three of them, Malaya, Kenya and Cyprus, and then drawn some valuable conclusions from the analysis which he has made.

Just as our knowledge and experience of conventional war have evolved over the centuries, there is now a parallel fund of new experience on which to draw for future counter-insurgency campaigning. But most of the authoritative writing on the subject has come from our opponents, such as Mao Tse-tung, General Giap and General Grivas; Colonel Paget helps to fill this deficiency in a very clear, interesting and well-argued book.

Beginning with Malaya and continuing with Kenya and Cyprus, he indicates the military problems facing the civil and military commanders and studies the steps they took to solve them, as well as analysing their methods of command and control. He shows how the triumvirate of 'civil, military and police' set to work, and how completely they relied on Intelligence to enable them to operate effectively.

He is rightly insistent on the necessity for the closest co-ordination and joint action by all the elements involved in any counter-insurgency. For centuries we have experienced a lack of co-operation between soldiers and civilians as well as between the Services themselves, and this was only rectified in the last war. Colonel Paget demonstrates clearly that successful counter-insurgency operations are dependent on a joint effort by all elements involved, working under one overall commander, either military or civil.

This book emphasizes that the end in this type of warfare must always be a political one; it is not for the soldier to decide the future of Ruritania—that is a matter for the statesmen. But the soldier expects to be given, and is entitled to demand, a clear political directive, which must include, and in fact be built around, the object which the civil Government wishes to be attained. The directive will inevitably involve the civil administration and also the police; once the police are called upon to act outside their normal duties, they will require the support of the military. So the triumvirate is formed of the civil administration, police and soldiers, all working together to attain the political aim.

This situation is not a new one. The British Army has been engaged in giving aid to the civil power for centuries, and the lack of publicity that it has received is a measure of its success. What is new is the probability of this situation continuing beyond the foreseeable future, and also the glaring publicity under which the Security Forces now have to operate. It is these factors which have given world importance to what, half a century ago, would have received little recognition, and no encouragement outside its local environs.

Today the circumstances are very different. A country which finds the normal diplomatic procedures too ponderous or ineffective can seek to achieve any aggressive aims by instigating insurgency instead of hazarding all on a declaration of war. It will need a considerable issue to provoke a nuclear conflict, but insurgency can always be labelled as a 'struggle for freedom', and can be promoted, financed and encouraged with a minimum of risk. Publicity can be used by the sponsors of the revolt to portray the insurgents as patriots and sufferers under the oppression of brutal and powerful 'imperialists'; and there is no lack of publicity today for such a cause. That is why Colonel Paget insists so rightly on adequate public relations coverage for our own cause in the early stages of any counter-insurgency campaign; it is much easier to create a well-informed public opinion from the start rather than to correct an unfavourable image which has already been projected to the world by the opposition.

When I was Director of Army Training, it was my task to train the Army for the next war; this presupposes that one knows the form which the next war can be expected to take. But unless we are going to start it ourselves, we shall have to conform, at least in the early stages, to the pattern initiated by our opponents, and they have a wide choice ranging from nuclear war down to plain terrorism. Based on the pattern of the last twenty years, it seems highly probable that insurgency in one form or another will play a part in most future campaigns in which the British Army is involved. We must therefore study counter-insurgency as seriously as we have in the past studied nuclear and conventional warfare, for it is a type of warfare which has come to stay.

Colonel Paget's book is a valuable contribution to current military thought on this important subject, and I commend it not only to all members of the 'civil, military and police' triumvirate, who may be involved at any time in counter-insurgency, but also to all who are in any way interested in the conduct of the cold war and the keeping of the peace.

PREFACE

In 1965 I found myself in Aden in a staff appointment directly concerned with the planning of measures, both civil and military, to be taken to defeat the insurgents then operating in those parts. The problems that arose were remarkably diverse and complex, but they were seldom completely new; they had almost all cropped up before in some previous Emergency, such as Palestine, Kenya, Cyprus or Malaya, and it would have been most helpful to be able to study this past experience and learn from it.

An excellent Joint Services pamphlet has now been compiled on counter-insurgency, but it was not available then, and there seemed few sources that could be tapped. So I began to devote my siesta hours to making my own research into past insurgencies, in an endeavour to find an answer to the many unresolved riddles piling up in my In tray. It proved to be such an absorbing subject that my notes eventually grew into this book, which I hope may help others who find themselves involved in counter-insurgency to produce better solutions than I ever managed and with rather less effort.

I would like to express my very sincere thanks to those who have helped me in the writing of my book. Field Marshal Harding and General Sir Gerald Lathbury have both most kindly given me of their personal experience and wisdom; Otto Heilbrunn, a great authority on guerrilla warfare, has also given me valuable comment and advice. I am particularly grateful to Brigadier Pat Hayward for his constant encouragement and for reading proofs and pointing out errors; also to the staff of the Ministry of Defence and Army Library for their assistance over research.

Julian Paget

Chapter 1

COLD WAR CAMPAIGNING

During the last twenty years, the British Army has been nominally at peace, and yet it has been involved in more than thirty operations of war, ranging from limited wars in Korea and Suez to 'policing' operations in Hong Kong and Jamaica. At times, there have been three major campaigns in progress at the same moment, and at the time of writing (1966) British forces are deployed in no less than eleven 'active service' or potentially operational areas.[1] There has in fact been no time since 1945 when British troops have not been involved in operations somewhere in the world, as can be seen from *Appendix A*, which shows every campaign involving the British Army between 1945 and 1966.

The conflicting 'peacetime' commitments of our Army today are clearly demonstrated by the ceremony of Trooping the Colour in London on the occasion of the official Birthday of The Queen. It might be thought that the traditional pageantry of this ceremonial parade was as far removed from active service as it was possible to be. Yet in fact, at least half of the 1,200 men on parade will either be overseas on operational tasks within a year of the parade, or else they will have returned from abroad within the previous twelve months.

In June 1966 for example, one Guards battalion taking part in the parade had returned from Aden the previous October, and another was under orders to move to that theatre the next

[1] Germany. Berlin. Cyprus. British Guiana. British Honduras. Aden. South Arabia. Bahrain. Hong Kong. Borneo. Malaysia.

October. Among the troops lining the Mall was the Guards Independent Parachute Company, which had until recently been in Malaysia, carrying out deep penetration patrols along the Indonesian border. Many of the Household Cavalry troopers on parade would, before the end of the year, be changing their chargers for armoured cars in either Germany or Malaysia. Yet another Guards battalion would soon be faced with two tasks in addition to its ceremonial duties; it had to prepare to move in October to Germany, where it would train for both nuclear and large-scale conventional war in a completely new role as an Armoured Personnel Carrier Battalion. At the same time, they would, while still in the United Kingdom, be required at times to be at seven days' notice or less to move anywhere in the world as part of the national Strategic Reserve.

This triple task—calling as it does for a tunic and bearskin in one locker, combat clothing in another and jungle green in yet a third—illustrates the problem of cold war campaigning as far as Britain is concerned. Although theoretically it is peacetime, we must be prepared for both nuclear and conventional warfare and at the same time be able to carry out limited war campaigns, counter-insurgency and internal security operations worldwide.

It may be helpful at this stage to make some definitions, in order to avoid confusion, since so many different terms have been used to describe the forms of warfare that have occurred during the last twenty years. The various cold war campaigns have been referred to as 'irregular', 'insurgent', 'partisan', 'guerrilla' and 'unconventional' warfare; the enemy have been called 'bandits', 'rebels', 'dissidents', 'guerrillas', 'insurgents' and 'terrorists'.

In this book, 'insurgency' is used to describe a form of armed rebellion against the Government, in which the rebels have the support or acquiescence of a substantial part of the populace; the methods that they adopt to achieve their aim of overthrowing the Government may include guerrilla warfare, but insurgents may equally well resort to civil disobedience, sabotage or terrorist tactics. Operations taken by the Government to defeat such a rebellion are 'counter-insurgency' operations; the military and police forces engaged in such a campaign are termed the 'Security Forces'.

'Guerrilla warfare' is used here in the sense of the special form of warfare, based on mobile tactics by small, lightly armed groups, who aim to harass their opponents rather than to defeat them in open battle. 'Internal security' operations, otherwise known as 'duties in aid of the Civil Power', are not to be confused with 'counter-insurgency'; they are aimed at controlling civil disturbances and are based on the principle of the use of minimum force, which does not apply in actions against insurgents.

The opposition throughout this book will be referred to as 'insurgents' if they are engaged in insurgency, 'guerrillas' if they are fighting a guerrilla war and 'terrorists' if they are indulging in terrorism.

An analysis of the time spent by the British Army on various types of operations since 1945 shows that the total period spent on active operations amounts to about forty years, and that of this, only five years or so have been devoted to conventional warfare as against thirty-five years or more which have been spent on counter-insurgency campaigns. These figures do not include the numerous 'internal security' operations which have also arisen.[1]

Nor is Britain alone in this commitment, for the United States, France and other countries have also been involved in varying degrees in similar counter-insurgency campaigns, as can be seen in *Appendix B*.

Insurgency has emerged as a science in its own right since the end of World War Two; it took the form during that conflict of guerrilla warfare and extensive sabotage by national partisans and Resistance Movements, but their activities tended always to be subservient to the main conventional warfare campaign. Since 1945, however, it has developed as a popular new technique for the attainment of political aims by subversive means in these days of so-called peace. It has been used by the Communists as a weapon in the cold war, and also by the leaders of

[1] See Appendix A.

numerous rebel movements as a means of achieving their aims, be they political, nationalist, racial or subversive. It has usually included both guerrilla warfare and terrorism among its methods and has, in almost every case, been directed at the overthrow, for one reason or another, of the authorized Government.

Insurgents cannot be fought successfully by conventional warfare methods, as the French discovered in their campaign in Indo-China 1946–54, and as the Americans are slowly learning again, over ten years later. As in all forms of warfare, the correct strategy and tactics can only be developed through a thorough knowledge and understanding of the enemy, his strengths and weaknesses, his psychology and his methods. This applies in particular to guerrilla and counter-guerrilla warfare, and the subject must be given the recognition which it deserves.

The continuing insurgent campaigns being waged against the Western Powers are a threat in themselves, but it is important to realize that they are a further threat in that they form part of the cold war. Post-war insurgencies have arisen from a variety of motives and in a variety of countries; but in many cases, they have been exploited to some extent by the Communist leaders to further their own aims. There is no evidence that the two Communist Powers have abandoned their long-term aims of achieving Communist domination of as much of the world as possible, and there is every indication that Communist China, at least, is becoming increasingly ambitious and aggressive.

Insurgent uprisings since 1945 have not necessarily been Communist inspired or controlled; the Emergencies in Palestine, Kenya and Cyprus all arose from other causes, as did those in Algeria, the Yemen and Aden. Some have been actively organized by Moscow or Peking however, with a view to overthrowing the existing Government and replacing it by a Communist régime, and these include China itself, the Philippines, Vietnam, Laos and Malaya.

Campaigns to defeat insurgencies require a considerable effort by the Government and the Armed Forces if they are to be successful; the campaign is usually lengthy and costly, particularly where guerrilla warfare develops, since it is a characteristic of this type of warfare that a handful of skilled guerrillas can tie down a totally disproportionate number of Security Forces for a long time.

Counter-insurgency campaigns are costly in money as well as in man-power, for not only are the actual operations a financial burden, (the French calculated that the Algerian War cost them £1 million a day), but the cost in terms of lost production and damage to economic interests can also be considerable. For example, in the Kenya Emergency the country's agriculture suffered and in Malaya there was damage to the rubber plantations and tin mines.

There is no easy or quick answer as to how to defeat insurgents, but much can be learned from a study of the many campaigns that have occurred in the past; they provide useful experience and lessons, which can be applied to the present and the future. We must not only profit from our own past knowledge, but we must try to anticipate possible future foes and their methods.

The Communists believe firmly in revolution as a means of overthrowing the existing systems, and their leaders have stated openly that they support revolutionary movements wherever they occur—unless, of course, they happen to be against themselves as in Hungary! They may give only moral support and training to the insurgents, they may supply arms and equipment, or they may resort to direct intervention, whichever seems appropriate. The 'uncommitted nations' of the world, and in particular, the emergent or newly independent ex-colonial territories are a primary target for Communist take-over bids, since they provide ideal conditions for anti-Western propaganda and are as yet inexperienced in countering the guiles of Communism.

This constant effort by the Communists to spread their creed by all means short of general war is referred to by the Free World as the 'cold war'; the Communists call it 'peaceful co-existence'. It is a conflict fought in the political, economic and ideological spheres as well as by military means, and these aspects of the struggle must be understood by any counter-insurgency forces, since they have a bearing on the military campaigns. We are concerned here primarily with the military problems, and will consider the other aspects only in so far as they affect the military issues.

The Communists in both Russia and China have made a detailed study of guerrilla warfare and as a result have developed a strategy for military campaigns in the cold war, based on three principles: the threat of military force, warfare by proxy and guerrilla warfare.

The Soviet Army is nearly two million strong, but it has never been in action since 1945, except to suppress internal revolts behind the Iron Curtain. Its very existence is used however by its masters as a threat to the West and as such, it is sufficient to tie down considerable Allied forces, particularly in Europe.

The Chinese Communist Army has been more aggressively employed and has fought in Korea, Tibet and India. Here again, even when it is not directly involved, the constant threat of its overt or covert intervention in the cold war in South East Asia is a significant factor. It encourages insurgents to fight on when they might otherwise consider accepting defeat, and it imposes caution on the West, both militarily and politically.

Most of the Communist cold war campaigning has been carried out by proxy; they have stirred up trouble and incited revolt wherever possible, the Communist Armies remaining as a threat in the background. Local insurgents have been used by the Communists to fight their battles for them, and only if they are not available at the right moment are the local Communist Parties thrown into the battle, as in Greece and Malaya.

Insurgency clearly has many advantages for this warfare by proxy. Firstly, the campaign can be fought as an 'undeclared war', with no overt commitment for the Communist Powers. Secondly, it enables them to carry the cold war militarily far beyond their own frontiers, simply by supporting other insurgent movements with arms and equipment and even on occasions, with 'volunteers'. Thirdly, it enables them to make the most effective use of their own supporters in any country in which they want to stir up trouble. Above all, insurgency is the most economical technique for their cold war campaigning. It is traditionally the method of the few fighting the many, and it costs the counter-insurgents far more in terms of money, manpower and effort than it does their enemies.

It is not difficult to see why the Communists seem to prefer cold war methods for the moment to a 'hot war'; while this is so, it must mean that the Free World, and Britain in particular, will continue to be faced by counter-insurgency campaigns, such as have formed most of our post-war operations. It is a basic rule that one should try to 'know your enemy', and the next chapter tries therefore to analyse the methods of insurgents and their organization.

Chapter 2

HOW THE INSURGENTS FIGHT

Insurgency may consist of guerrilla warfare, sabotage, subversion and terrorism, or only some of them, according to the circumstances. This chapter analyses the organization and the aims of modern insurgent forces, as well as the essential requirements on which they will base their campaign. Finally, it considers the strategy that might be adopted by future insurgent leaders.

ORGANIZATION

Insurgent forces usually contain both a military and a political element, and, as with conventional forces, the former are normally subordinate to the latter; this is certainly the case where the Communists are in charge, and any Communist-inspired insurrection will be closely controlled by a tight command structure. Other rebel movements may work along similar lines, but probably with a looser and less elaborate hierarchy. A typical insurgent organization is illustrated at *Appendix C* together with an outline of the organizations that were adopted by the insurgents in Malaya, Kenya and Cyprus.

In overall control will be a political leader, who may, or may not, also be the military commander. He will preside over a *Central Committee* where general policy will be decided; in a Communist structure, there may also be an 'inner cabinet' of three or four members only, called the Politbureau.

The political structure at lower levels will depend on the

nature and the size of the country concerned. There will normally be committees in each Region, State, and District, and they will in turn be divided into several Branches based very roughly on each village or settlement.

The Branch Committees are of considerable importance, because they are responsible for maintaining contact with and controlling the 'Masses', who are the uncommitted or sympathetic local populace. The Branches have the task of indoctrinating the masses and organizing them to give support to the active insurgent forces.

A special committee which is formed when appropriate, and usually at State level, is a Border Committee, designed to organize operations and sanctuaries in the area of any interstate or international boundary.

The insurgent military chain of command runs parallel to the political structure, and will be headed by a military commander, with his own *Military Headquarters*, working in cooperation with the senior political committee. Below him, there will normally be a *Regional Headquarters* with the responsibility for controlling a number of subordinate units within its area.

The basic unit of any guerrilla force is usually a 'group' of 10–30 men, which may be called a gang, platoon, cell or some similar title; it will operate independently in an area allotted to it, and will be controlled directly by the Region with no intermediate headquarters. Large units are not kept permanently in being, since this is incompatible with guerrilla tactics; they are formed as required, by concentrating several 'groups' for a particular operation, and then dispersing them again.

These 'groups' are the cutting edge of the insurgents' sword, and they form the 'active' wing with varied tasks according to the type of campaign being fought. Where the terrain is suitable for pure guerrilla tactics in the form of 'hit-and-run' attacks, ambushes and raids, the guerrilla groups dominate and are in the majority. In urban conditions, as in the towns of Cyprus and in Aden, other types of groups are required, such as 'Killer groups', bomb-throwers or saboteurs, in order to implement a policy of terrorism.

These 'professional' groups are not to be confused with the 'amateurs', who are often students and schoolchildren, with the function of couriers, food suppliers, recruiters and producers of pamphlets. The latter are part of the 'passive' wing, and are probably under the control of the Branch Committees rather than of the military leaders.

This system of small groups is forced on the insurgents by their need for dispersion and mobility, and it suffers from the resultant weakness that effective control and good communications are both difficult to maintain. Messages usually have to be passed by courier, and are slow and uncertain; junior commanders are consequently very much isolated and find themselves unsupported when things go wrong. The system is not therefore very flexible, and is liable to collapse if attacked or undermined by the Security Forces.

Security is always a problem to the insurgents who have to guard constantly against subversion, informers, traitors and deserters. Their security is always of the highest standard, and vital information is protected by adopting the 'cell' system, whereby only one person in each group knows details of future plans and can identify his next superiors and juniors. This ensures that each individual has very little to give away to the enemy, even if he tells them all he knows. But on the other hand, it can be turned against them by skilful propaganda, or by penetration of the organization.

AIM

The aim of any insurgent force is to achieve a political objective by military means. It is rare, though not unknown, for the insurgents to achieve a decisive military victory by defeating the Security Forces in battle. Military success is not however essential to them, for they can still gain their objective, if they can enforce a stalemate, in which the Security Forces finally decide that the cost of continuing an indecisive struggle is no longer worthwhile. Their purpose is to defeat the Government's will to fight on, and military action is only one factor in their struggle to achieve this.

BASIC REQUIREMENTS

In order to conduct a successful campaign, an insurgent force needs to be assured of the following basic requirements:

(*a*) A cause to fight for.
(*b*) Support from the local populace.
(*c*) Bases.
(*d*) Mobility.
(*e*) Supplies.

Conversely, of course, if the Security Forces can deprive the insurgents of any of these essentials, it may prove to be a more effective method of defeating the revolt than hunting individual rebels.

A Cause

The life of an insurgent is usually a hard one, particularly when things are not going favourably. He will be constantly hunted and harassed by patrols, harangued by propaganda, certainly uncomfortable and perhaps hungry. His morale and determination to fight on will depend to a considerable extent on having a worthwhile cause to fight for. This cause will need to be simple, inspiring, and convincing; Communism, for example, may sometimes be inspiring enough in itself, but it is more effective if presented as the overthrow of the 'wicked imperialists' in a particular country, and the setting up of a revolutionary régime, in which the insurgents will be privileged members. The régime will, of course, be intended as a Communist state, but this particular pill will be well coated by assurances that the cause is 'nationalism', 'freedom' or some such popular slogan.

The cause must be convincing to the local populace as well as to the guerrillas. Enosis, for example, met with considerable support from the Greek-Cypriots, who represented 80 per cent of the island's population, and the insurgents therefore had a valid 'popular cause'. The Communist cry in Malaya of 'down with British Imperialism' did not however seem a worthwhile objective to the Malays, and since they formed half the population, this was one of the major reasons why the revolt failed.

Nationalism has always been a popular motive for rebellion, as is 'social injustice' or oppression; these causes all have the advantage too that they appeal to the populace as well as to the insurgents, and so tend to bring the two closer together, thus assuring the insurgents of their second need, the support of the local population.

Support of the Local Population

Any insurgent organization is dependent on the local population for several necessities, such as food, information, recruits and above all, for freedom of movement; the insurgent must be able to rely on the people not to betray him, even if they do not actively support him. He must, to repeat Mao Tse-tung's apt simile, 'be able to live among them as a fish lives in water'.[1]

If the insurgents have the active and willing support of the people, as in Palestine and Cyprus, the task of the Security Forces is made infinitely more difficult. Conversely, if the people defy the rebels and co-operate with the Government, as in the later phases in Malaya and Kenya, the insurgency cannot long survive. To continue the analogy of a fish in water, if the Government can transfer the sympathy of the populace from the insurgents over to themselves, it is equivalent to draining the pond, and so removing the environment in which the insurgents have been living. How the insurgent seeks to gain and to retain this vital support is discussed later in this chapter.

Bases

Particularly in the early stages of any insurgency, the rebels must have bases from which to operate, especially if they are fighting a guerrilla campaign. If the terrain is suitable, the bases will be set up in the country concerned, (as in the forests of

[1] *Selected Works of Mao Tse-tung*. 1954.

Kenya and Malaya, and the Troodos Mountains of Cyprus). They will usually be small, carefully concealed and well dispersed; no attempt will normally be made to defend them, the survival of the guerrillas being considered more important than the retention of territory.

The ideal position for the guerrillas is to be in possession not only of secure bases, but also of an inviolate 'sanctuary', accessible to them, but not to the Security Forces, because it is situated across an international border in a neighbouring territory which is sympathetic to the guerrilla cause. In Greece, the ELAS members were able to withdraw into Communist Yugoslavia and Albania, and the Algerian rebels of 1954–63 fled into Tunisia when pressed; the guerrillas in South Arabia now are based and trained in neighbouring Yemen, and only infiltrate into South Arabia to carry out attacks, before returning to complete safety.[1]

Supplies

Although their logistic requirements are few, guerrillas still need arms and equipment, food, medical supplies, money and clothing. Here again, the ideal is to have an outside source, such as the Yemen, but a sympathetic local population is almost as effective, except that it is exposed to counter-measures by the Security Forces.

Mobility

Guerrilla tactics demand that the tactical offensive is always in guerrilla hands, and this in turn depends upon their ability to achieve and to retain the initiative; the guerrillas must be free to attack the Security Forces on terms of their own choosing and to escape without undue casualties. Such 'hit and run'[2] tactics depend on the ability to move freely to and from the scene of the battle, to concentrate and then to disperse again. Guerrillas

[1] See Appendix B.

[2] 'Thousands of small victories eventually turn into one great success' (General Giap).

seldom attempt to retain ground, because to do so places the initiative in the hands of the Security Forces, and also deprives the guerrillas of their greatest advantage, mobility.

In undeveloped country, this need for mobility is usually met by the use of inaccessible terrain such as forest or mountains, where the guerrillas have space and so security. In urban areas, they achieve their freedom of movement by mingling with the local populace and not being recognizable.

STRATEGY

Mao Tse-tung laid down a strategy for large-scale guerrilla warfare, such as he himself fought in China,[1] but this does not necessarily apply to the form of guerrilla warfare likely to be adopted by insurgents such as Mau Mau, EOKA and the Communists in Malaya. Although the tactics may be much the same in each case, the strategy differs, primarily because Mao Tse-tung based his strategy on a protracted campaign, whereas insurgents seek a quick decision, if possible.

The leaders of an insurgency must hold on to their great advantage of surprise and the initiative, and will normally try to use it to achieve their aim by a sudden coup, thus avoiding the need for a full-scale campaign. This was attempted in Malaya in 1948, but failed, and the Communists were forced into a twelve-year war—which then also failed.

Even if the coup is unsuccessful, the insurgents will still hold the initiative for a time, and must use it to the full to win a decision, before the Government can recover and muster its full resources. During this phase, the insurgents will establish bases from which they will resort to guerrilla tactics to gain control of certain populated areas, from which they can then expand their influence. Their success or failure will depend on the speed and vigour with which the Security Forces react to deprive them of the initiative.

If this second opportunity is also lost by the insurgents, they are then committed to a protracted campaign, and this calls for

[1] *Selected Works of Mao Tse tung*, 1954.

a change of policy. They cannot at this stage hold any great hope of defeating the Security Forces in open battle, and they will not usually attempt to do so. They will aim instead at forcing a stalemate by preventing the Government from defeating them, and by aiming to weaken the Government's will to fight on.

To achieve these aims, they will adopt true guerrilla tactics with a view to tying down the maximum number of troops and causing as much nuisance, expense and embarrassment as possible to the Government. They will play the part of a vicious gnat stinging and eluding a larger, rather clumsy beast, until it retreats in fury and frustration.

The Communist leaders have studied the theory of guerrilla warfare very thoroughly, and have also gained first-hand experience. The Russians established a highly successful partisan movement against the German invaders of their country in World War Two, and sponsored a civil war in Greece 1944–45. The Chinese under the veteran guerrilla leader, Mao Tse-tung, fought the Nationalist Armies for twenty-two years before finally defeating them in 1949.

Mao Tse-tung's theory of guerrilla warfare, based on his personal experience, has been studied and put into practice by other guerrilla leaders, such as Giap, Chin Peng and Ho Chi Minh. Furthermore, his teachings are accepted as gospel to the extent that they are being passed on to potential insurgents in other countries worldwide, and they deserve therefore to be treated seriously, as semi-official doctrine, which may be put into practice by future insurgents.

A further aim of the insurgents during any campaign will be to rally public sympathy, both locally and if possible worldwide, to their support, for this can be a decisive factor in their favour. The Cyprus Emergency provides a good example of the skilful manipulation of world opinion by the Cypriot and Greek leaders. This aspect of any insurgency is of ever-increasing significance today and its importance must not be underestimated by the Security Forces.

A final factor which the insurgents cannot afford to ignore is the support of the populace in the territory in which the campaign is being fought. Their strategy depends upon it, and they must not only ensure that they obtain this vital support from the start but also that, having gained it, they do not then take any action which might lead them to forfeit it.

Part of the population in any country involved in an insurgency will usually be in sympathy with the insurgents, and part will be firmly opposed to them. The majority, however, will be passive and uncommitted, sitting on the fence until they can see which side offers the best long-term prospects and the best security now and in the future.

The insurgent will attempt to win preferably the active support, or failing that the acquiescence, of the populace by the application of the 'carrot and stick' technique. Initially, there will be subtle indoctrination of 'the masses' by propaganda and psychological warfare methods. False promises and distortions of the truth will deceive some into turning against the Government; threats will discourage others from co-operation with the civil authorities and the Security Forces; extortion and blackmail will force some into unwilling acquiescence in the insurgent aims.

Once the campaign starts and the authorities begin to take their own measures to counter the insurgent activities, the latter are forced to resort to stronger measures themselves. Then it is that intimidation begins, following a pattern that is all too familiar. Loyal citizens are tortured or killed, their houses are burnt and their property damaged. Indiscriminate attacks are made on crowds, and the Police and Special Branch are primary targets. Above all, 'informers', 'collaborators', 'traitors' and 'spies' are publicly threatened and then murdered. Schoolchildren are indoctrinated into support of the rebel cause as a patriotic duty, and parents acquiesce to protect their children and themselves.

Intimidation, ruthlessly applied, is a powerful and effective weapon in the insurgent's armoury. But it can be a double-edged weapon, and all insurgents need always to ensure that it does not turn against them, as it may do if it is employed clumsily or to excess.

In Cyprus, EOKA killed more Greek-Cypriots than they did

British and Turks, but they retained their grip on the populace. The Mau Mau also killed many more of their own tribe than they did Europeans, but they went too far when they perpetrated the Lari massacre,[1] which turned public opinion irrevocably against them.

Another technique to gain local support is the administration of oaths of loyalty to the insurgent cause; the oath itself gives the rebels some control over those who take it, and this can be effectively strengthened by some judicious murders if the oath is broken.

Finally, the insurgents can adopt the subtler techniques of trying to provoke the Security Forces into taking security measures which are repressive or unpopular, and which thus tend to swing the sympathies of the public against the Government and therefore towards the insurgent cause.

The outcome of the final phase may be a military success for the insurgents, as in Indo-China and in Cuba. If it is not, they may nevertheless gain a favourable political outcome by forcing a stalemate. Alternatively, they may themselves be defeated militarily, as in Malaya and in Kenya, or they may be defeated politically, as in Cyprus, when the insurgents were ordered by their own political leader to cease fighting.

Having had a look at how the insurgents fight, and how a campaign might develop from their point of view, it will be of interest now to study the form that it is likely to take, as seen by the Government and the Security Forces against whom it is waged.

[1] See p. 93.

Chapter 3

THE PATTERN OF INSURGENCY

General

A definite pattern exists in the military aspects of any counter-insurgency campaign, and it is both interesting and of value to analyse it and to see what can be learned from it. The pattern falls into three distinct phases, which can be applied to most operations in which the British Army has been involved since 1945.

Phase One forms the prelude to military operations, and consists of a period of increasing lawlessness by some element of the population which is in conflict with the Government. This dissension leads in due course to the stage where the civil authorities can no longer cope with the situation on their own and are forced to call for military assistance to enable them to restore law and order.

Phase Two sees the initiation of military action against the insurgent organization concerned. But it is essentially a defensive phase, in which the insurgents hold the initiative, while the civil, military and police forces are hurriedly built up, and preparations are made for offensive action as soon as possible.

Phase Three is the final stage where the initiative passes to the Security Forces, who then bring increasing pressure to bear on the insurgents, until it is considered possible to end the Emergency and to seek a political solution to the problem.

It is worth considering these three phases in more detail.

PHASE ONE

The opening phase usually starts with some element of the

populace concerned having a grievance, real, contrived or imagined. This develops into a clash of interests between them and the Government, in which the insurgents decide sooner or later that the time has come to achieve their aims by violence rather than by peaceful means. The cause of the trouble may be the thwarting of political ambitions, as in Cyprus; it may be purely Communist subversion as in Malaya or it may be a tribal revolt, as in Kenya.

These rebellions may or may not be Communist inspired, but even where they are not, they are nevertheless 'viewed with favour' by Moscow and Peking, because they almost certainly further the cause of Communism by embarrassing and weakening the 'capitalist' power against whom the revolt is directed.

During Phase One it is still the responsibility of the Civil Administration to maintain law and order, and we see the authorities trying to control the situation by normal peacetime methods; but effective action is hampered by laws and procedures which were not intended or designed to cope with the subversive activities now facing them. The Police, and Special Branch in particular, find their peacetime resources becoming increasingly strained by the additional burdens being thrust upon them.

Any military forces in the area will also probably be on a peacetime basis, stationed there either as a garrison or in order to use that territory as a base for operations elsewhere. The headquarters will be geared primarily to administrative rather than to operational duties, and any interest that the troops may have in local security will normally be limited to being prepared to disperse riotous crowds as part of their 'duties in aid of the civil power'.[1]

The pattern of insurgency follows a remarkably predictable course in this first phase. The discontented element will usually be banded together as a secret organization, be it Communist,

[1] 'The British Army in Cyprus, owing to the nature of its organisation, was a clumsy weapon, completely lacking in the training necessary for this kind of fighting. The same applied to the police. . . .' (*Guerrilla Warfare* by General Grivas, p. 8.)

EOKA or Mau Mau, plotting and planning their campaign. They will be well organized, but as yet probably poorly equipped; they will have excellent security, a fierce belief in their cause and a detailed plan of campaign; they will hold the initiative, in that they can achieve surprise in deciding when to start active operations. One of their chief weapons at this stage is intimidation, which is usually directed primarily against their own people, in order to ensure their support for the revolt, however reluctant and however passive. They will indulge in limited propaganda and carry out some careful recruiting; they will most likely have some comparatively innocent semi-political organization, which they will use as a 'Front' for their more sinister activities.

As their campaign against the authorities builds up, they will resort to increasingly drastic measures, such as murders and violence, in order to intimidate the populace and to neutralize the Police or Special Branch; there may be sabotage or raids on armouries to acquire weapons and explosives; there will also be the elimination of selected individuals branded as 'traitors' or 'collaborationists'; at the same time the hardcore of guerrillas will be organized and prepared for the fight, while the political element will be organizing the 'support wing' among their followers in the populated areas.

Once it becomes evident to the authorities that this is the work of a subversive organization against which the existing resources of the Administration are unlikely to prove adequate, the need for Emergency measures begins to be considered, and, in particular, powers to detain suspects without trial. But the authorities will be reluctant to take any more drastic steps than are essential, for it is an admission of failure to have, for example, to resort to repressive measures and to call for military support. The moment will come, however, when their inability to deal with the situation can no longer be ignored or concealed, but must be publicly acknowledged by the declaration of a State of Emergency.

PHASE TWO

The taking of this decisive step usually brings a sense of relief, like the bell to signal the opening round of a fight; but it is not always apparent at this stage how long the contest is likely to last, and some unduly optimistic forecasts are all too often offered.

This is a depressing phase for both the civil and the military authorities, because the initiative almost inevitably lies as yet with the opposition. This is the moment for which the insurgents have long been planning and striving, but which the authorities had hoped would never come; as a result, the Government and consequently the Security Forces are usually unprepared and are thrown on to the defensive until they can complete their preparations, which they could not or would not initiate until the Emergency was declared.

Additional military forces will be flown in and an appropriate command organization established. Initially the troops will, of necessity, be dispersed on static guards and other duties in support of the hard-pressed police and local forces. But as soon as possible they will be re-grouped to form a striking force, however small, which can take offensive action against the insurgents as the first step towards regaining the initiative. Full use will be made of the local forces, who know the country and speak the language, to advise the newcomers on the techniques most appropriate to the area.

The Police Force and Special Branch will need considerable strengthening. This will be done by calling up Reserves, enlisting Special Police and also recruiting from outside the territory. Wherever reinforcements are brought into the Police and the Civil Administration, it requires time for them to fit in and to master the local situation. Great care is required at this stage to integrate the 'new broom' with the 'old hand', so as to make the best use of both.

An effective civil-military command structure is essential to enable a proper policy to be developed at the highest level, and to ensure that executive orders and decisions are passed downwards without delay. This structure usually takes the form of a *War Council* or *War Cabinet*, composed basically of the civil and military leaders, together with Police, Intelligence and other representatives as required. This close civil-military co-operation continues downwards from Government House level, through Provinces and Districts to the 'front line'. At each level in the command structure, the battle is planned and fought by the well-tried triumvirate of 'civil, military and police', which has prosecuted the cold war in a score of campaigns throughout the world during the last twenty years, or more, and will doubtless continue to do so for many years to come.

The relationship between the Governor and the Commander-in-Chief (or their equivalents) with regard to the control of the Emergency is largely dependent on the scale of disruption to the normal administration of the territory. In Cyprus and in Malaya the whole country was involved in the Emergency, and the two posts were therefore combined, for the peak of the campaign, under one person who had full control over all aspects of life in the country, both civil and military. In Kenya, the Mau Mau were confined to an area which was only one-sixteenth of the whole colony; the Governor therefore remained in overall control of the Civil Administration, and the Commander-in-Chief was responsible for the security operations.

There is sometimes a demand for the declaration of Martial Law in the trouble area, as being the most effective means of eliminating terrorism. This is an extreme step, however, which is fortunately seldom, if ever, necessary; it is justified only when the civil administration has collapsed completely, and the military commanders have to assume control. Due no doubt to the ghost of Oliver Cromwell, lurking still in the lobbies of Westminster, Martial Law is not even provided for under the law of the land, and the military commander who takes it upon himself to declare it can only rely afterwards upon being granted an Act of Indemnity rather than his bowler hat!

Special legal powers are a different matter and are recognized as being essential in the peculiar circumstances of an Emergency, if the Security Forces are not to be impossibly handicapped in their efforts to defeat insurgents who operate outside the law. This requirement is met by Emergency Regulations, which give the Government and the Security Forces very much wider powers than would be acceptable under normal conditions. Nevertheless the legal position is invariably a subject of much emotion and controversy between the supporters of two opposite points of view. There are those who demand exceptionally stern punishments for the insurgents and wide powers for the Security Forces, on the grounds that ruthless terrorists are otherwise allowed by the niceties of the law to escape the punishment they deserve, and also that innocent lives are endangered for the same reason. On the other side are those who oppose any deviation from normal procedure, on the grounds that the good name of British justice rates higher than any short-term advantage, and that nothing must be allowed to undermine the tradition of British insistence on absolute justice. Clearly, there is a middle path, but it is not always easy to find.

The relationship between the Police and the Armed Forces is that the latter normally take over responsibility for internal security from the police once the Emergency is declared, but this is not an invariable rule, and the military may continue to act 'in support' of the police even during the Emergency. This depends largely on the threat to security, the relative operational effectiveness of the two forces, and the allocation of duties between them; and perhaps on personalities! There can be no hard and fast rule, but in general, where troops are employed in a military rather than a police role, they should be in control of the operations.

The behaviour of the local population in an Emergency is a vital factor and it follows a very standard pattern. The actual insurgent element of the population is nearly always very small, less than 10 per cent, but it succeeds in instilling such fear into the majority of the people that the remaining 90 per cent almost invariably side passively, if not actively, with the insurgent cause.

This is understandable for several reasons. Firstly, the populace not unnaturally want to be on the winning side, and indeed it is very much in their interest that they should be; they therefore try to offend neither side, until they can see which way the struggle looks like going. Secondly, they are certainly afraid to support the Security Forces against the terrorists unless they can be protected against insurgent reprisals; intimidation is a most unpleasant, and often fatal, experience. The Mau Mau proved its effectiveness, when they succeeded in forcing loyal African servants of many years standing to turn against their European masters, the alternative being to see their own families tortured and perhaps killed. Finally, the sympathies of the majority of the population may well lie with the insurgent cause, even if they do not agree entirely with the methods used to attain it, as in the Cyprus Emergency.

The formation of some form of Home Guard is one of the measures which are open to the Security Forces during Phase Two, in order to gain control over the insurgents. It not only fortifies the morale of the loyalists, but also helps to isolate them from the subversive elements, an important factor in the defeat of the revolt.

A second measure, which not only helps to isolate the insurgents, but also serves to free troops from static duties, is 're-settlement' or 'villagization', as it is sometimes termed. This consists of moving groups of isolated and scattered locals into larger, specially prepared settlements and 'new villages'; on their own, these small dispersed groups are not only extremely difficult to protect against attack, but they provide the enemy, either voluntarily or under duress, with valuable food, shelter and information. It is a hard decision to uproot families from their homes, but in the long term, it has proved to be to their advantage in both Malaya and Kenya, where it was implemented; it not only gave them protection against intimidation, but also made it possible for the Government to provide them with far greater amenities and benefits, such as schools, roads, shops and hospitals, than was possible when they were scattered round the country. This measure cannot be implemented in urban conditions such as in Cyprus.

The military aim during this second phase will be to move over to the offensive as soon as possible, but this cannot be achieved until the necessary build-up of forces is complete, and enough has been learned of the opposition to form the basis of effective operations. As the Police are reinforced and the Home Guard grows, the Armed Forces will try to hand over complete areas to the Civil Administration, and so free sufficient troops to form a striking force. Meanwhile, military operations will be aimed at accustoming troops to the special local conditions, to harassing the insurgents in any way possible and to gaining information for future use.

It is not always easy to assess the moment at which Phase Two moves into Phase Three. But there comes a stage where the Security Forces can feel that they are beginning to gain the upper hand against the insurgents and are able to impose their will on them to some extent. They will have achieved some public successes against them, and this may have induced some of the waverers to begin to question whether the insurgent cause is perhaps the best one to follow after all. This is a crucial moment and may well be the turning point in the campaign.

But the beginning of the end of the military campaign does not mean the solution of the problem by a long way. The struggle will be continuing meanwhile in the field of politics, diplomacy and propaganda, not only within the territory concerned, but also outside in the world and in the United Nations. Local military success will not necessarily lead to success in these fields.

PHASE THREE

For the Security Forces, the final phase is rather like the month of March, in that 'it comes in like a lion and goes out like a lamb'. It sees them emerging from the frustrations of Phase Two, and embarking on the last lap, with a smell of victory in the air. Successes against the insurgents become more frequent, until it is more a matter of hunting down the surviving remnants, who by now no longer represent any real threat to security, but are only concerned to save their own skins. Finally, the task of the Security Forces is completed, and the ultimate solution of the problem is left in the hands of the political leaders.

During this final phase, the population may or may not shift their loyalty towards the Government, depending largely on the political background to the struggle; in Malaya and Kenya, the population swung round to the Government, but in Cyprus they favoured Enosis to the end, rather than co-operate with the Government for the sake of peace without Enosis.

Truces and surrender offers are usually proposed towards the end of any conflict, and a counter-insurgency campaign is no exception; here again, the differing views of the 'oppressors' and the 'appeasers' emerge. The former maintain that it is wrong, with victory in sight, to relax at all, but that on the contrary, pressure should now be built up to eliminate those terrorists who were themselves so ruthless when things were going their way. On the other side, are the opponents of 'unconditional surrender', who remember 1945; they look ahead beyond the end of the Emergency to the need for goodwill in the future rather than for a more complete military victory in the present.

Political factors come more and more into prominence at this stage, and military action is increasingly adjusted accordingly. As we noted earlier, the cold war is a struggle for the hearts and minds of the people, and this will continue long after the military conflict has been concluded, regardless perhaps of how successful it has been. When the fighting ends, the military commanders and their troops depart for other battlefields, but the Police and the Civil Administration remain, continuing the struggle to build up something constructive and lasting from the ashes, blood and bitterness left behind. The elimination of the insurgents does not necessarily eliminate the basic conflict of interests which led to the revolt; it only makes it possible to negotiate under rather more favourable conditions.

The pattern of cold war operations is exemplified to my mind by the role of the triumvirate of the 'civil, military and police'. They form, as it were, a skilled and dedicated medical team, who assemble as necessary to work together on a patient who is suffering from some form of insidious poisoning. The Civil Administration play the part of the hospital authorities, watching over the health of the patient as one member of their community. The Commissioner of Police is more like the family doctor (he is certainly comparably dedicated and overworked!). He will know his patients personally, having watched over them for many years; he is therefore likely to be the first one to spot any unhealthy symptoms, be it contagious Communism, the EOKA itch or the fever of nationalism. He will report the symptoms to the hospital authorities, though he may on occasions be told he is being 'alarmist'. The illness will meanwhile be spreading, infecting the bloodstream and leading to outbreaks of festering sores.

Eventually, the malady will prove to be beyond the powers of both the doctors and the hospital authorities, and they will be forced to admit that the only solution is to call in the surgeon in the form of the Armed Forces. But he will need time to prepare his instruments, and to confirm the diagnosis before he operates; and throughout the operation, he will bear constantly in mind that he must not cut too drastically, lest he prejudice the future health of the patient.

The operation duly completed, the surgeon then removes his mask and gloves, and hurries off to another urgent case (though he will, of course, keep his first patient under careful observation for a while to ensure that the operation has succeeded). He will, however, only have removed the source of the infection, and it will remain for the doctors to treat the patient correctly from then on, to restore his health and to strengthen him against further infection. This will call for careful nursing and treatment, so that he can become a useful and a grateful citizen.

It is a simile which can be adapted and elaborated indefinitely, but it may be of interest to apply it now to some of the counter-insurgency campaigns in which the British Army has been engaged. The three selected for this book are the Emergencies in Malaya (1948–60), Kenya (1952–56) and Cyprus (1955–59); each has its own features and its own lessons, and all of them can help us to plan for future counter-insurgency campaigning.

The Emergency in Malaya
(1948-60)

Chapter 4

MALAYA – PHASE ONE (1945–48)

When British officers of Force 136 were landed behind the Japanese lines in Malaya in 1943 to fight a guerrilla campaign against the invaders, they found a well-organized guerrilla force already in existence. It consisted of some 6,000 Chinese Malays who called themselves the Malayan People's Anti-Japanese Army (MPAJA). One of the leading figures of this organization was a clever young man of twenty-one, called Chin Peng, who co-operated so effectively with the Allies that he was later chosen to march with them in the Victory Parade in London; and a year after that, he was awarded the Order of the British Empire for his services to the Allied cause.

The Investiture did not however take place. When the decoration arrived in Malaya in 1948 to be presented by the High Commissioner, Chin Peng was once again hidden deep in the jungle; but this time he was acting in his capacity of Secretary General of the Malayan Communist Party (MCP) and was busy preparing for a Communist revolt against British rule. His award was duly cancelled and was replaced not long afterwards by a reward of £30,000 for the capture of Chin Peng, the insurgent leader.

This phase in the life of Chin Peng mirrors that of the Malayan Communist Party, and both have a very direct bearing on the development of the Emergency in Malaya. The MCP was originally formed in 1927 at the instigation of Moscow; it consisted almost entirely of Chinese, for the peaceful, happy-go-lucky Malays were not interested in violence or in revolution, and were quite content with British rule. When Japan invaded the country in February 1942, it was the Chinese Communists who undertook, in agreement with the British Government, to form a guerrilla force that could continue the struggle in Malaya. The MPAJA was formed therefore and was the only Allied force behind the enemy lines until December 1943, when it linked up with the advance elements of Force 136.

As the end of the war came into sight, the MPAJA increasingly turned its attention from military action to the revival of Communist political ambitions. Its leaders began to look ahead to the post-war phase, and plans were laid to take advantage of the confusion that was likely to follow the defeat of the Japanese, by setting up a Communist régime in Malaya before the British Government could re-establish its authority. This plot was fortunately thwarted by the setting up of the British Military Administration, which established effective control of the country just in time.

In December 1945 the MPAJA was officially disbanded. Some 6,000 guerrillas handed in their arms and accepted their two campaign medals and gratuity of £45. In theory they returned to civilian life, but in fact they promptly joined the officially recognized MPAJA Ex-Services Comrades Association, which was a thinly disguised 'front' for what was virtually a Communist Reserve Army in Malaya. A further 4,000 MPAJA members never emerged at all, but remained in hiding in the jungle, preparing for the moment when they would be called upon to raise the banner of revolt against the 'British imperialists'.

Malaya in the immediate post-war years must have seemed to the Communists to be as ripe for a take-over bid as it would ever be. Firstly, it was a territory still under colonial rule and therefore suitable ground for nationalist propaganda. Secondly, the 'wind of change' was blowing already in South East Asia, and could conveniently be exploited by the Communists to replace Western rule by régimes more favourable to Communist purposes. Thirdly, the power and prestige of the Western Powers had suffered an incalculable reverse as a result of their humiliating defeat at the hands of the Japanese. Finally, while the Allies were reducing the size of their Armed Forces as rapidly as they could, the Communists retained their covert guerrilla forces as well as their highly organized, officially recognized Communist Parties, so as to be ready to attack the unsuspecting 'colonialist powers' as soon as the moment seemed propitious.

Conditions in Malaya are highly favourable for a guerrilla war.[1] Four-fifths of its 50,600 square miles (slightly less than England and Wales) is primeval jungle with a woven canopy of treetops through which the sun cannot penetrate, while underneath, a matted mass of bamboo and undergrowth reduces movement in places to an exhausting 200 yards in an hour. Down the backbone of the country runs a 6,000-foot mountain range with endless ridges and streams branching from it. The jungle is largely impenetrable except by animal tracks or by hacking a path through the undergrowth. The country's 3 million acres of rubber plantations and 700-odd tin mines, as well as its rice fields, lie almost entirely along the west coast, which is very much more highly developed than the east, and contains the majority of the 12,000 European population.[2] Malaya produces one-third of the world's rubber and tin, a rich prize that the Communists would dearly like to see transferred from the pockets of the West to their own.

Strategically too, the Malayan peninsula forms, in the hands of the West, a barrier to Communist expansion into the Indian Ocean, while under Communist control, it could become a springboard for further Communist advances westwards.

The Civil Power in the country was the Federal Government, headed by the British High Commissioner, then Sir Edward Gent. He was advised by an Executive Council, equivalent to a Cabinet, while the Parliament consisted of a Legislative Council of seventy-four members. (See *Appendix D*, column *b*.)

Politically, the country was divided in 1948 into eleven separate governments—nine Malay States and the two Settlements of Penang and Malacca. Each State had its own Ruler, who was

[1] See Map p. 79.

[2] Population figures, 1948: Malays 2,600,000; Chinese 2,000,000; Indians 578,000; Europeans 12,000.

head of its Government and had a certain amount of autonomy. Under him was a Malay 'Mentri Besar' (Prime Minister) and a State Council with a Malay majority. The two Settlements were administered by British officers of the Malayan Civil Service. The administration of the country was primarily, however, in the hands of the strong Central Government.

The Malayan Communist Party was at this time an officially accepted political party, but they were busy using this unusual state of grace to prepare for the subversion of the Government. Their activities followed the usual pattern of infiltration of the Trade Unions, the establishment of 'cells' in every village, intensive indoctrination of the people wherever possible, and intimidation of key individuals who neglected to show adequate enthusiasm. As usual, they formed a Communist youth organization, and so began to nurture further recruits.

Their activities first became evident in 1947, when there were 300 major industrial strikes. A general strike in Singapore led 173,000 workers to down tools. In September, attacks were made on several planters and it became all too clear that law and order were being threatened.

Then in February 1948, there was a meeting in Calcutta of Communist leaders from Russia and all the South East Asian countries, under the pretext of attending the World Youth Festival. It is probable that the decision was here taken that a policy of violence should be adopted in order to further the spread of Communism in South East Asia; this is not conclusive, but it is hard to accept as a pure coincidence the fact that, within six months of this Communist gathering, Communist-inspired insurrections broke out almost simultaneously in Indonesia, the Philippines, Burma and Malaya.

In Malaya, industrial unrest, the intimidation of officials and labour, as well as attacks on persons and property increased sharply in the months following the conference, and the Police and Intelligence reported that they believed it to be the deliberate work of a Communist-inspired subversive organization. On 1st June 1948 there was a general demand that Sir Edward Gent should take strong action, and ban the Malayan Communist

Party. He did not do so, but on 9th June the police were given powers to deport anyone found guilty of intimidation or violence.

On 16th June, three European planters were callously murdered in their offices by gangs of armed Chinese, one of whom declared, 'We will shoot all Europeans.' Two Chinese working for Europeans were also shot. Angry planters demanded immediate action, and that afternoon, the High Commissioner declared an Emergency in parts of Perak and Johore States only.

There was still considerable disquiet, however, and demands that stronger action be taken; on 17th June 1948, the High Commissioner declared a State of Emergency throughout the Federation of Malaya.

Chapter 5

MALAYA—PHASE TWO (1948–51)

The Emergency in Malaya was to last twelve years, before the revolt was finally crushed. The campaign is of particular interest, because it arose from a deliberate Communist act of aggression, rather than being a national insurgency, which was merely exploited by the Communist leaders. It is the only instance we have of a direct guerrilla war between Britain and the Communists themselves, and it is therefore of the greatest value to study the organization, strategy and tactics of both sides, particularly as the Communists were finally decisively defeated.

The declaration of the Emergency brought no immediate relief; the initiative remained with the Communists, and within two weeks they had shown their power. Well-armed gangs of guerrillas attacked two Police Stations in Johore and one in Pahang; on 12th July, another gang attacked Malaya's only coalmine, Batu Arang, twenty-six miles from Kuala Lumpur, killed five people and occupied the township of some 12,000 people for an hour. With the element of surprise on their side, the 4–5,000 Communist guerrillas could and did strike fiercely as and when they wanted; they operated in groups of 100–200, which were strong enough to overrun isolated police posts, and there was little that the Government could do.

On 29th June, the High Commissioner flew back to London for consultations, and four days later, he was killed in an air crash. His successor, whose appointment was not announced till September, was Sir Henry Gurney, former Chief Secretary to the Palestine Government.

The police, however, did have one lucky break, which was to help them considerably. On 16th July, a special police squad of fourteen Chinese and Malays under the command of Inspector Bill Stafford closed in at dawn on a hut near Kajang in Selangor in which it was believed that two not particularly significant Communists were hiding. Three men rushed out, of whom two were killed and one escaped; the hut was then searched, and found to contain maps, four rifles, three shotguns, three pistols, a sack of ammunition—and six Chinese women.

As the police were sorting out their trophies and burning the hut, they were suddenly counter-attacked by thirty to fifty guerrillas and a fierce gun battle followed, in which six guerrillas and five of the six captured women were killed, with no police casualties.

One of the first two guerrillas killed turned out to be Lau Yew, who was the commander of the whole guerrilla Army and a member of the Central Military Committee of the Malayan Communist Party; the one woman who survived the battle was his wife and she had escaped in the confusion. Lau Yew had been presiding at a top-level meeting called to plan the capture and occupation of Kajang in order that it could become the guerrilla headquarters. The members at the meeting had each brought their own bodyguard of five to ten men, and it was they who had launched the counter-attack. This unexpectedly successful police operation not only thwarted what might have been a considerable coup for the guerrillas, but inflicted a severe setback to their plans and their morale, and must be regarded as an important incident.

The Malayan Communist Party was officially proscribed by the Government on 23rd July, but by then it had been reorganized on its insurgent basis deep in the jungle, with the title of the Anti-British Army. This was changed in February 1949 to the 'Malayan Races Liberation Army' (MRLA) in an attempt to indicate the nationalist nature of the revolt; but in fact, the insurgents were 90 per cent Chinese, and the Malays showed no inclination to join on any scale.

The MCP in the jungle was planned on an elaborate structure

of committees and military units.[1] The controlling body was the *Central Committee*, headed by Chin Peng, the Secretary-General, and thought to number ten to fifteen altogether. Hidden deep in the jungle, close to the sanctuary of the Thailand border, it met once or twice a year only, to issue policy directives which filtered very gradually down to the front-line fighters. The three leading figures formed the Inner Cabinet as it were, called the Politburo, and they were valued by the Government together at some £60,000.[2]

Below the Central Committee was a widening hierarchy, consisting of *Regional Bureaux* (one for North and one for South Malaya), *State Committees, District Committees* and *Branch Committees*; the Branches were responsible for controlling the Min Yuen, who were 'the Masses', who had to be organized to help and support the active insurgents.

The Min Yuen were a vital element of the MRLA and formed their logistic support, intelligence agencies and Public Relations Branch. They consisted of the local Chinese still living in the villages who, either willingly or out of fear, supplied the jungle gangs with food, money, information, recruits and couriers as called for.

As such, the Min Yuen were from the start made a priority target for the Security Forces, for if they could be disorganized or prevented from fulfilling their role the jungle gangs would be crippled. It was always a principle of the Malayan Government to attack the political and support organizations of the insurgents, for not only were they more accessible than the guerrillas, but they were the centre of the insurgency. To pursue our fishy similes, it was like attacking the head of an octopus rather than the individual tentacles, and it did prove extremely successful.

The MRLA had a total strength of 6–8,000, and was organized into Regiments of 200–400 on a basis of one or more to each

[1] Appendix C, columns (*c*)–(*d*) and Appendix D, column (*e*).

[2] Chin Peng, Secretary-General, £30,000; Yeung Kwo, military leader, £15,000; Lau Lee, propaganda chief, £15,000.

State.[1] They were sometimes divided into Companies but more often into Independent Platoons of fifteen to thirty, which, in the early stages, joined up for operations, but as pressure against them grew, they remained scattered, and operated in isolation, each in its own area. They consisted almost entirely of guerrillas, but they did carry out sabotage when opportunity offered, and they indulged in widespread extortion and intimidation. There was some terrorism, but it was secondary to the guerrilla warfare.

The MRLA lived a hard life, lightened only by the prospect of the day of victory. For the first year or so, they occupied camps for up to 600 men, concealed deep in the jungle, well built and with some degree of comfort. But as the campaign progressed and the gangs were steadily broken up, these became smaller and rougher. The guerrillas followed a disciplined routine, with Reveille at 0530 hours, followed by two hours of drill before breakfast. The rest of the day probably included two hours of political indoctrination, three hours of fatigues, and a closing two-hour session of that Communist privilege of 'self-criticism'!

Discipline was stern, with the death penalty for many offences, and flogging up to thirty-six strokes. Visits home were allowed for the first year, but then stopped, when 'absence without leave' increased.

The British Army in Malaya in 1948 was under the command of Major-General Sir Charles Boucher, and consisted of three British battalions, six Gurkha battalions, and two Malay Regiment battalions, supported by one field regiment and a Malay coast artillery unit. In August, another British battalion arrived from Hong Kong, and in October, 2nd Guards Brigade consisting of three Guards battalions arrived from the United Kingdom. It was the first time that Guardsmen had served in Malaya, but

[1]

	Regiments
Johore	3, 4, 9
Pahang	6, 7, 10 (Malay)
Perak	5
Negri Sembilan	2
Selangor	1
Kedah	8
Kelantan	Pulai Gang, then 7 Coy of 5 Regt.

despite being welcomed by an American correspondent with the words: 'Well, Britain's pride of the palaces are just one more bunch of boys in the basha-huts of Malaya . . .'[1] they took to the jungle fast and soon established a fine fighting reputation.

A less orthodox formation was Ferret Force, which was set up as an *ad hoc* measure; it consisted of ex-Force 136 members, British, Gurkhas and Malays, all skilled in guerrilla warfare and possessing a thorough knowledge of the country. They penetrated in small patrols deep into the jungle and stayed there for weeks, hunting down the Communist gangs. They were disbanded after a few months only, and dispersed as guides and instructors among the Regular Army units, but in that time they had demonstrated a technique of deep patrols that was to be generally adopted by the Security Forces in due course, and was to prove highly successful.

The Police Force, with a strength of only just over 10,000 in 1948, was desperately hard pressed, and urgent steps had to be taken to strengthen it. Several hundred members of the former Palestine Police were transferred, and a call for Special Constables produced the fine response of 30,000 Malay volunteers within six months. They were then formed into 'Area Security Units', each some thirty-five strong, and were of considerable value for static guards, patrols, road checks and searches.

The Regular Police Force also formed some 200 'Jungle Squads' which fought alongside the Army, particularly in the setting up and manning of 'jungle forts' in the aborigine territory.[2] A Police Frontier Force was also formed to patrol the border with Thailand; the guerrillas had been moving freely across the border and using Thai territory to rest and reorganize, but this was curtailed considerably when Thailand agreed in 1949 to allow Malayan police to pursue guerrillas for up to ten miles into Thailand.

The new Commissioner of Police, appointed in September 1948, was Colonel W. N. Gray, C.M.G., D.S.O., who had been Inspector-General of Police in Palestine, a strong and resolute

[1] *Menace in Malaya*, by H. Miller.

[2] See p. 72.

man, who did a very great deal to organize and train the rapidly expanding Malayan Police Force.

The Security Forces did not make the mistake of remaining on the defensive, but took the offensive as far as possible. It was a matter at this stage of counter-attacking rather than attacking, and the technique adopted was to station a strong force of near company strength in a central position between a group of villages. From there, they aimed to be on the scene of any guerrilla attack within thirty to forty minutes, and then to pursue the guerrillas relentlessly. This worked remarkably well, and gradually drove the Communists to give up large-scale operations and to break up into smaller, less formidable groups of only fifteen to twenty.

One decisive civil measure that was taken was the compulsory registration of the population. All citizens over twelve had to be in possession of an identity card bearing their photograph and fingerprint. The Communists quickly appreciated the danger to them of such a step, and therefore did everything they could to sabotage it. They initiated propaganda asserting that it was only being implemented by the authorities for sinister reasons such as to conscript men for the Army, to collect more taxes or to requisition property and food. They called it 'tyrannous enforcement of fascist methods', and threatened death to the officials who organized it, and even to the photographers involved. When this had little effect, they indulged in a few judicious murders. They also destroyed every identity card that they could lay their hands on, raiding villages and ambushing buses in order to do so. Despite the intimidation and the propaganda campaign against registration, the measure was pushed through, and life was made more difficult for the Communists throughout the country from that moment.

Other Emergency Legislation gave the Government power to detain suspects for up to two years without trial, to control movement, to impose curfews and to search without a warrant. There were also severe penalties for aiding the insurgents in any way, and a mandatory death penalty was imposed for carrying arms and ammunition.

One somewhat controversial measure was Regulation 17D which gave the High Commissioner the power to order the detention of all the citizens of any village or area proved to have been aiding the guerrillas. It was harsh, and also distasteful, in that it was a collective punishment affecting innocent as well as guilty. It was invoked sixteen times during the first nine months of 1949 and some 6,000 people were detained; once the resettlement of squatters was carried out, it became less necessary and was then only rarely used.

By mid-1949, there was evidence that the Communists were becoming discouraged at the lack of success of their early ambitions. The people had not rallied to their support as expected, the guerrillas had been forced on to the defensive and no revolution had been achieved. By August, the incident figure was down to twenty-six per month, the lowest since the Emergency began.

Encouraged by these signs, the Federal Government tried out a surrender scheme, and on 6th September dropped a million leaflets over the jungle, declaring that, although men known to be guilty of murder would still be tried by law, and could be hanged, those who surrendered voluntarily would not suffer the death penalty, then in force, for the illegal possession of firearms.

It did not produce the hoped-for wave of surrenders, and brought in only 116 insurgents by the end of the year. The Communist leaders, however, reacted sharply by making the handling of the leaflets punishable by death and by tightening their discipline generally.

They also launched a new offensive, and on 11th September a force of no less than 300 guerrillas attacked Kuala Krau, a railway town in Pahang, killing four police, two British civilians and two Malay women. They derailed an armoured train and burned the station as well as other buildings. Several other attacks by gangs of 100–200 also occurred, and the incident rate rose sharply during the last quarter of 1949.

Nor did 1950 start any better. The number of incidents continued to rise alarmingly, and the British Government did not help, when in January, it announced its recognition of the Communist Chinese People's Republic. This move greatly encouraged the insurgents and was equally discouraging to the loyal elements in the country, both Malay and Chinese, who needed much convincing thereafter that the Government was really still serious about fighting the Communists in Malaya, while apparently supporting them in China.

In February 1950, guerrillas burned the village of Simpang Tiga in Perak, leaving 1,000 people homeless; they also attacked a Police Station at Bukit Kepong in Johore, killing twenty-three police and civilians. In May, the number of incidents reached the grim total of 534 for the month, twice the previous highest monthly total.

The outlook was not encouraging. Little progress had been made in loosening the grip of the insurgents on the local population; it was not as yet possible to protect them against Communist infiltration and intimidation, and they were therefore understandably reluctant to oppose the insurgents by failing to supply them with food and information when ordered. Only the European planters and miners refused steadfastly to be cowed, despite the very considerable risk and strain that they faced; indeed, production of both rubber and tin actually rose in 1950—rubber to its second highest total ever.

There was by now increasing dissatisfaction in the country over the conduct of the Emergency, and it was not without justification. The main weakness was the absence of any one person responsible for the handling of the Emergency as a whole, so that the civil, military and police effort could be co-ordinated in one effective policy. The Army were still acting 'in support of the Civil Power' and they felt unduly hampered by the meticulous methods of the Police and the cautious attitude of the Civil Administration; at the same time, some Government officials, for their part, felt that the military showed a certain lack of appreciation of the civil problems.

The command and control position was also unsatisfactory, in that there was no clearcut division of responsibility between the Army and the Police; this inevitably led to a clash of viewpoints between soldiers who saw their task as the defeat of the insurgents by all available means within the shortest possible time, while the police had to consider the legal and civil implications of any action taken, and so might seem rather overcautious.

It was appreciated that drastic action was required, and several moves were made. 26 Gurkha Infantry Brigade moved to Malaya from Hong Kong, and was followed by 3 Commando. Additional aircraft were allotted, including a squadron of heavy Lincoln bombers from Australia. Most significant of all, Lieutenant-General Sir Harold Briggs was appointed as Director of Operations 'to plan, to co-ordinate and to direct the anti-bandit operations of the police and fighting forces'.

General Briggs had commanded the famous Fifth Indian Division during the Burma Campaign and had been Commander-in-Chief in Burma until he retired in 1948 and went to live in Cyprus. At the personal request of his former commander, Field Marshal Slim, then Chief of the Imperial General Staff, he agreed to take on the job of Director of Operations in Malaya for eighteen months, working as a civilian directly under Sir Henry Gurney.

A quiet and conscientious man, he devoted himself wholeheartedly to the formidable task that faced him, so much so that when he left Malaya, he was sick and worn out, and died only eleven months later. During his tour of office, he suffered many frustrations and never managed to gain the executive powers from the Government that he considered to be essential, and which he had expected to be given. But he was a man who got things done whatever the difficulties, and he initiated, and left to be completed, the 'Briggs Plan', which was nothing less than the Master Plan for the winning of the campaign.

The Plan was issued within a fortnight of his arrival, after he had made a swift tour of the country, talking to all kinds of people. It set out four main aims:

(*a*) To dominate the populated areas and to build up a feeling of complete security, which would in time result in a steady and increasing flow of information coming from all sources.

(*b*) To break up the Communist organizations within the populated areas.

(*c*) To isolate the bandits from their food and supply organizations in the populated areas.
(*d*) To destroy the bandits by forcing them to attack the Security Forces on their own ground.

The first step towards achieving these aims, and perhaps the most significant step, was the integration of the Civil Administration, the Armed Forces and the Police into one co-ordinated team, working to a common plan in close co-operation rather than in isolation. To emphasize this 'joint approach' to the Emergency, he set up the first truly joint command and control structure to direct security operations. (*Appendix D*, column (*d*),

At the Federal Government level, an *Emergency Operations Council* (EOC) was set up, headed by the Prime Minister, and including in its members the Ministers directly concerned with the Emergency, the Director of Operations, the Service chiefs and the Commissioner of Police. This body was equivalent to a War Cabinet and decided overall policy for the conduct of the Emergency. Under the EOC, General Briggs had his own *Director of Operations' Committee* (DOC) consisting of himself, the Service Chiefs, the Commissioner of Police and a representative of the Minister of Defence. This small committee met regularly at least weekly and took policy decisions on the use of the Security Forces within the wide policy agreed by the EOC.

A similar system was also established within the States and Districts, each of which had its own policy-making *War Executive Committee* (WEC) with the appropriate civil, military and police members. Again, within each WEC, an *Operations Committee* was set up to be responsible for the day-to-day executive control of security operations. In each committee, it was a matter of co-operation and the pooling of ideas, 'discussion and decision', with every member contributing and none dictating.

The system of integrated control was completed by an insistence on complete military-police co-operation, exemplified by the establishment of joint military-police Operations Rooms in all areas. All intelligence and incident reports were channelled through them, and they were jointly manned by police, military and Special Branch staffs.

Another field in which this integration was particularly significant was Intelligence, where all resources and sources were pooled under one Chief of Intelligence, and an integrated organization was built up.

The principle of joint control of the Emergency was thus firmly established, and a machine set up by means of which joint decisions were taken and then implemented without argument or delay. This organization was not only to survive the duration of the Emergency with virtually no change, but was to be the pattern for future counter-insurgency campaigns.

The strategy now, in order to achieve the aims set out in the directive, was to isolate the populace from the guerrillas and so deprive the latter of several of their essential needs. The first step was to reinforce the Police and Special Branch and set them to attack the Communists and their political organization in the populated areas.

At the same time, the troops were ordered to form secure bases in the villages, and from them to clear and to dominate the surrounding jungle 'up to about five hours journey'. This technique served to deter and to disperse the gangs and to drive them deeper into the jungle, where they were less of a menace; nor were they allowed to rest by retreating into hiding, for deep patrols and air attacks hunted them out and harassed them even there, destroying their food and arms dumps and burning their camps.

A policy of 'food denial' was also instituted, which was in the long term to contribute largely to the defeat of the guerrillas; the police efforts against the Communists in the villages did something to control the steady flow of supplies to the guerrillas, but it was virtually impossible to prevent them from using the thousands of Chinese 'squatters' as an alternative source when things became more difficult in the villages. From this sprang the imaginative and far-reaching plan of 'resettlement'.

The 'squatters' were the thousands of Chinese who had for one reason or another established little plots of agriculture along the fringes of the jungle; they were widely scattered and therefore highly vulnerable to pressure and intimidation from the guerrillas. They helped the jungle gangs in many ways besides providing food, for they could also be called upon for money, information and recruits; they could not be distinguished from loyal citizens and so formed a dangerous Fifth Column, which the Administration found it difficult to control. General Briggs saw the value that could be gained by moving the 'squatters'; the Government had already prepared a White Paper on the problem and agreed to his proposals.

There were some half a million 'squatters', but by the end of 1951, 400,000 of them had been moved into some 500 'new villages' at a cost of £11 million. The bare statistics give little idea of the magnitude of the task; land had to be found in suitable areas, and the States Governments had to be persuaded to allot it to suspect Chinese squatters rather than to Malays. Labour forces had to be reorganized so that rubber plantations and tin mines were not deprived of vital workers. Roads had to be built, communications provided, building material found and Chinese-speaking administrators appointed. Above all, the confidence of the people themselves had to be won, for this was not only a security measure, but also a major social reform which could have a great bearing on the future of the country. This was made more difficult by strenuous Communist efforts to prevent and to sabotage the operation, which they saw clearly would be to their disadvantage.

But the resettlement went through, helped by the troops themselves; one group of 'squatters' were so grateful to the Coldstream Guardsmen for their cheerful aid, that they asked for, and received, permission to name their new home 'Kampong Coldstream'. Each family ended up with one-sixth of an acre on which to build a hut and grow food, as well as a cash allowance to cover their needs for five months; schools and shops appeared, and a community life developed.

It was certainly a success as a social measure, and it was of great value operationally, for as a result, the guerrilla bands were forced to move or to split up still farther afield in order to obtain adequate supplies. More important, it enabled the Security

Forces to extend the isolation of the guerrillas, by giving the populace a degree of protection which was previously impossible, and so removing them from the grip of the insurgents.

The year 1950 ended with a new spirit of enthusiasm and hope in the country; both the people and the Security Forces felt that they had a plan to follow and were achieving results. Indeed, the insurgents had lost some of the initiative, and were not operating in such large groups or on the same scale as a year before. As they were driven farther from the populated areas, so the troops followed them up, and adopted small-scale tactics of patrols and penetration of the deep jungle to hunt them down.

1951 did not unfortunately maintain the progress made in 1950. Despite the ascendancy gained by the Security Forces, the insurgents launched a fresh offensive in the New Year, and the rate of incidents climbed steadily to a new high level in June of 606 in the month. It was found that the Communists had changed their tactics and were infiltrating among the labourers on the estates and among the remaining squatters, and extorting food from them still.

Two steps were then taken to counter this development. First, the labour forces were regrouped into areas where they could be controlled and guarded, and the cost of this move was placed on the employers, despite some strong protests. Secondly, drastic new restrictions were imposed to enforce still further the policy of food denial; it was called 'Operation Starvation' and was to be a determined effort to cut off the guerrillas from all sources of food, medicines and other essentials. All areas from which the enemy might possibly obtain supplies were declared to be 'controlled areas', from which it was forbidden to take food, money or clothing; shopkeepers had to keep detailed records of goods sold and the names of all their customers; vehicles carrying food had to be covered and lashed down and were forbidden to halt during any journey, so as to prevent the surreptitious delivery of supplies to jungle gangs. Rice was rationed and issued already cooked, in which state it remained edible for only two or three days. The hope was that if the people could tell the Communists with truth that they had been completely prevented from aiding them, this might help to defeat the Communist extortion techniques.

It was a harsh restriction, meaning that a man could not even take any food with him when he went out for his day's work. It was also not easy to enforce, requiring considerable manpower to ensure that the rules were obeyed. The State of Negri Sembilan achieved this by a remarkable effort of co-operation called 'Operation Hubbard'; the Police, Army, civilians and Government officials all played their part, and senior officials including the British Adviser and the Malay State Secretary took their turns manning road blocks on the outskirts of the capital, Seremban.

As part of 'Operation Starvation', General Briggs called for the organizing of Home Guards, to assist in enforcing the regulations and to free the police for other tasks. But there were few arms to give them and the Government was reluctant to issue what there were, for fear that they would find their way into enemy hands. It was to be some time yet before this move was to get going.

On Saturday, 7th October 1951, the High Commissioner was assassinated. As his Rolls Royce, escorted by two police vehicles, approached an S-bend on the road from Kuala Lumpur to Fraser's Hill, sixty miles north, it came under concentrated fire from the high banks on either side. As if to draw attention away from the car, Sir Henry Gurney got out, closed the door behind him and walked away; he was shot dead within a few yards. The car was hit thirty-five times, but miraculously, Lady Gurney, the Secretary and the driver were all unhurt.

The death of Sir Henry Gurney brought not only deep anger, but also insistent demands for stronger action against the insurgents, for new policies and for a greater sense of urgency. The hopes of a year ago had faded, and the end of the Emergency now seemed further off than ever.

In November, General Briggs left Malaya, exhausted and disillusioned, unaware perhaps of just how much he had in fact contributed to the ultimate victory that he was not to live to see. He was replaced by General Sir Rob Lockhart.

On 29th November, the Secretary of State for the Colonies, Mr. Oliver Lyttelton, came out to Malay to see things for himself, and was forcibly advised that the overriding need was for a really strong and determined overall commander. Early in the New Year, the choice was announced as General Sir Gerald Templer, G.C.M.G., K.C.B., K.B.E., D.S.O., who had been C-in-C of Eastern Command since 1950. He was to be appointed High Commissioner with 'full and direct responsibility for the Emergency'. So that he would not be tied down by non-operational matters, he was to have a Deputy High Commissioner, responsible for all administration; this appointment was given to a young but extremely able civilian named Donald MacGillivray, formerly Colonial Secretary in Jamaica.

It was a precedent to appoint a serving officer as High Commissioner; it was never doubted that he would deal effectively with the military situation, but it was reasonable to question whether a professional soldier would be able to handle the complex political problems of Malaya, involving as they did the nine Malay States and the two Settlements, as well as the overriding problem of the Chinese population, on whose attitude depended to a great extent not only the outcome of the Emergency but also the future of Malaya.

It was a challenge which was to be fully met and mastered, and the arrival of this fresh leadership marked a new phase in the conduct of the Emergency. The campaign was to last another eight years, and there would be much grim and relentless fighting still ahead; but the initiative was henceforward to remain firmly in the hands of the Government, and despite the depressing record of 1951, there was now little likelihood of a Communist victory, short of outside intervention.

The insurgent loss of the initiative came to light in an unexpected way. Early in 1952, a Communist document was captured, dated September 1951, which turned out to be a directive by the Central Executive Party of the MCP itself, addressed to all State and District Committees. It did not accept defeat, but it did make the remarkable admission that the current Communist policy of murder, sabotage and intimidation was a mistake, in that it was losing the insurgents the support of the local population. They were therefore ordered to change their tactics, and to try to 'conciliate' the people, instead of terrorizing them. Unnecessary 'inconvenience' to the masses was to be avoided, attacks against the Security Forces were not to endanger the public and food collection was not to depend on intimidation; at the same time, everything possible was to be done to win over the populace to support of the insurgent cause. The Party members were also instructed to adopt a policy of infiltrating wherever possible into Government organizations, in order to exert an influence that way.

The directive took many months to reach some of the guerrilla bands, which explains the high rate of incidents early in 1952 and the decline in the second half of the year. It was not made public to the people of Malaya until October 1952, and they were unaware therefore during the depressing previous twelve months that the enemy were in fact equally depressed at the way things were going!

The beginning of 1952 therefore sees the Communists frustrated, while the Malayan authorities were slowly gathering confidence, as a new phase began under fresh leadership; the defensive period was over, which offers a suitable moment to end Phase Two of the Emergency.

Chapter 6

MALAYA—PHASE THREE (1951–60)

Phase Three began on an optimistic note. The capture of the Politbureau directive in 1952 had revealed that the Security Forces had wrested the initiative from the insurgents, even if they had not realized it at the time. The Briggs Plan had laid the foundations for the conduct of the campaign. And now, General Templer had brought to Malaya two battle-winning factors—an official promise of independence in the near future, and his own personality and military skill.

The decision to grant independence to Malaya in 1957 was of great significance, because it virtually won the 'battle for the hearts and minds of the people'. From that moment, the Government could rely on the support of the population, both Chinese and Malay, against the Communists; at the same time, the insurgents' promises of 'liberating Malaya from British imperialism' were now shown to be weightless and very much less attractive than the British Government's offer of self-determination.

The other decisive development was the arrival of General Templer himself. He landed on 7th February 1952, and was sworn in as High Commissioner the same day. That done, he immediately set out to meet as many people as possible, both official and unofficial, to hear their views and to make his own policy absolutely clear.

Wherever he went, the High Commissioner drove home four main points. Firstly, the conduct of the Emergency and normal civil Government were completely inter-related and would be handled as one problem. Secondly, the whole population must play their part in fighting the Communists. Thirdly, the insurgents must be defeated before Malaya could expect self-government. Finally, and most important of all, he asserted again and again: 'The answer lies not in pouring more troops into the jungle, but rests in the hearts and minds of the Malayan people.'[1]

He set about implementing his policy with an energy and a dynamic personality which spread an infectious enthusiasm and spirit of confidence throughout the country. His reorganization covered every field.

The command and control structure was streamlined to integrate civil, police and military still further; the Federal War Council was abolished and incorporated into the Federal Executive Council, which now became the High Commissioner's personal 'instrument of policy', planning both the civil and the military conduct of the campaign. He reorganized the military command structure, and himself took over the duties of Director of Operations, with General Lockhart as his Chief of Staff; he also set up his own Joint Operations Room in Kuala Lumpur, from which he controlled all security operations.

The efficiency of the Security Forces themselves was improved by the acquisition of more and larger helicopters, which were to prove a tremendous asset; more arms were also obtained and 12,000 were issued to the settlers. At the same time, the High Commissioner began to build up a Federal Malay Army, and gave new impetus to the forming of the Home Guard; the defences of the kampongs and the 'new villages' were improved, and additional measures were taken to protect road and rail traffic. The police were re-trained and given more and better equipment.

A clearcut policy for the Security Forces was laid down, based firstly on isolating and protecting the civil population from the insurgents, and secondly, on offensive operations. Alongside this, General Templer set out his long-term aim which was to defeat not only the insurgents, but Communism itself, by creating a new Malayan nation and a new Malayan way of life, where Communism would have no place. But he emphasized to the people

[1] *Templer in Malaya* by C. N. Parkinson, p. 23.

that this could be achieved only with their full support and co-operation; it could not be left to the British alone to defeat Communism.

The minds of the people had been largely won over, but it was another matter to win their active support against the insurgents, who were still able to intimidate the populace. It was, therefore, very rightly decided that the defeat of this terrorism was to be a priority task from now on, so that the people could be given 'freedom from fear'.

The Security Forces were now at their maximum strength. There were 40,000 troops in the theatre, composed of about 25,000 British, 10,000 Gurkhas and 5,000 from other Commonwealth countries.[1] In addition, the police were now some 40,000 strong.

The guerrilla strength could not be estimated exactly, but was thought to be between 6,000–8,000, which gave the Security Forces an apparent superiority of 15 to 1. In terms of man for man in the jungle, however, the numbers were much more closely balanced; infantry battalions in action against the guerrillas totalled 23,[1] and each had at most 400 men actually operating in the jungle, making a total of about 9,000 'bayonets'. Police support on these operations amounted to some 8,000 men, which gave the Security Forces a superiority of not more than 2½ to 1. This is of interest in the light of suggestions that the British Army is unable to fight guerrillas except on a basis of overwhelming superiority in numbers. Of the 40,000 troops in Malaya, approximately 30,000 were supporting or administrative elements.

In support of the Security Forces was the Home Guard. It had been formed initially in 1950, but had made slow progress. General Templer quickly realized the important part that the Home Guard could play, and built them up so that by 1954 they numbered nearly 200,000, and eventually reached a ceiling of 250,000. In 1953, he took a risk and issued 1 in 3 of the Home Guard with weapons for the first time; he even provided them with some armoured cars. The gamble paid off, few weapons were lost, and the Home Guard from then on played a significant part

[1] *Battalions:* British, 6; Gurkha, 7; Malay, 7; K.A.R., 2; Fiji, 1. Total, 23.

in the campaign. They formed a valuable link between the Security Forces and the populace, and they provided much useful information. They were able to improve the security of their own villages considerably, and thus freed trained troops from static duties.

The majority were used only on these static guard duties, but some volunteers were formed into Operational Sections of twelve men for duties in the jungle, usually attached to an infantry battalion. The majority of the Home Guard were Malays, but in the Kinta Valley, where there were Chinese-owned tin mines, the mine owners found their own predominantly Chinese Home Guard, and themselves paid half their cost.

As the numbers of the Home Guard grew, so did their problems, and in 1953, Major-General E. B. de Fonblanque (retired) was appointed as their Inspector General, with responsibility for their training and administration. Operationally, they were controlled by the State War Executive Committees through State Home Guard Officers (SHGO's), with an equivalent at District level.

The Home Guard helped to isolate the population both physically and mentally from the guerrillas; and this isolation was reinforced by increased police effort in the villages and rubber plantations. They now made a determined drive against extortion and intimidation, and set out to win the confidence of the people. One imaginative move was initiated in December 1952 by the new Commissioner of Police, Colonel Arthur Young of the City of London Police, who had replaced Colonel Gray in January 1950. He called it 'Operation Service' and began by sending a personal letter to every member of the Police Force with the instruction '... do everything you can ... to extend kindness and help to those in need'. It met with a remarkable response, and up to 20,000 instances of 'service' a month were reported; it was a major contribution to the battle for the hearts of the people.

Meanwhile, the policy of 'food denial' had been continuing and it was beginning to prove effective in certain areas; some surrendered guerrillas admitted that their rations were at times reduced to a cigarette-tin of rice a day, a story which was confirmed by their emaciated condition.[1] It was indeed becoming more and more difficult for them to obtain food from the populace, despite most ingenious smuggling methods, such as concealing rice in walking-sticks, bicycle-pumps, girls' brassières and underneath refuse carts. Special Branch had begun to infiltrate agents and informers into the Min Yuen, and they trapped many smugglers as a result. As the Communist Branches lost their members, replacements to continue this vital role were transferred from the jungle guerrillas; this reached such proportions that it was calculated that by 1957, 90 per cent of the surviving guerrillas were then in the Branches rather than fighting in the jungle gangs.

The main task of the Security Forces, however, was to carry out offensive operations against the guerrilla gangs in the jungle. Primarily, this consisted of endless patrolling to locate, pursue and harass the small bands lurking in the vast areas of jungle and swamp; it was a grim manhunt in country so close that men could stand feet apart and not see each other in the sunless, half light. It was a specialist type of warfare, but the British soldiers, 50 per cent of them National Servicemen, soon became highly skilled, although they could seldom equal the natural aptitude of the Gurkhas. They asked only one thing, which was to make contact with the enemy; but this was a rare satisfaction, for it was calculated in 1956 that it worked out at an average of 1,800 man-hours of 'jungle-bashing' for every 'contact'.[2] Nevertheless, by the end of the Emergency, the Security Forces had killed six guerrillas for every soldier or policeman killed, and had proved that the British Army could outfight the guerrillas on their own ground.

It was found that it paid to harry and pursue any guerrillas contacted as relentlessly as possible, driving them from one camp to another, forcing them to move and to disperse. Such methods lowered guerrilla morale and encouraged deserters, who otherwise had little incentive to surrender, if they were left in peace and comparative comfort.

[1] It was estimated that the minimum requirement was about 5 lb. of rice per man per day.

[2] One battalion calculated that they wore out 15,000 pairs of jungle boots in three years in Malaya—an average of 20 pairs per man.

Surrendered insurgents were handled by the police, and some of them were most successfully employed by Roy Sutcliffe, Head of Special Branch; he formed a Special Operations Volunteer Force, composed of surrendered Communists who were willing to return to the jungle and to lead patrols against their former comrades. They were paid at normal police rates and after eighteen months of service, were released unconditionally. They proved surprisingly willing 'turncoats' and helped the Security Forces considerably to penetrate and to break up the insurgent organization.

Considerable efforts were made to persuade the insurgents to surrender and to give information. Very large rewards were offered for the capture of the leaders or for information leading to their capture, the top price being £30,000 for Chin Peng himself, down to £250 for a non-ranking bandit. Ex-insurgents were eligible for 50 per cent of the rewards, and they gave away much valuable information; but there were also criticisms that former guerrillas could in this way earn enough to retire in comfort!

Psychological warfare was just beginning to get rather slowly into its stride, and in 1953, 93 million anti-Communist pamphlets were distributed, of which the RAF dropped 54 million—together with 3,500 tons of bombs. Voice aircraft, (Valettas fitted with loudspeakers), flew over the jungle from 1953 onwards, describing the security and comparative comforts of captivity as opposed to jungle life; the message was reinforced by dropping pictures of emaciated guerrillas on capture compared with their well-fed grins a few weeks later. Personal messages by surrendered insurgents were also broadcast, but the surrender rate remained low, for discipline in the 'platoons' was still rigid and effective.

These measures, together with the increased protection now being given to the populace, led to a slight but encouraging improvement at last in the flow of information and intelligence. This greatly increased the chances for the Security Forces to make contacts with the guerrillas, instead of being forced to comb the jungle in the faint hope of stumbling on the enemy.

It was still, however, a matter of fighting and working for every piece of intelligence; it came from captured insurgents and their documents, from patrols, air photos, and from the painstaking collection and collation of clues; but it still did not come in any quantity from the most valuable and bountiful source of all—the local population. This was dramatically demonstrated by the 'incident' at Tanjong Malim.

On 24th March 1952, the water pipeline supplying the 20,000 inhabitants of this town, fifty miles north of Kuala Lumpur, was sabotaged by guerrillas for the sixth time in four years. A repair party set out at 7 a.m. next day, escorted by a police jungle squad and accompanied by Michael Codner, the local Assistant District Officer; he was already well known as one of the three officers involved in the Wooden Horse Escape from a German prisoner-of-war camp, and had also distinguished himself by outstanding work as an administrator in Malaya. Two miles from the town, the party walked into a carefully laid ambush and there was a fierce battle in which Codner and eleven others were killed, eight wounded and only one member of the party survived uninjured; two out of an estimated forty guerrillas were killed, and the others fled when reinforcements appeared.

Three days later General Templer himself arrived in Tanjong Malim, and, summoning the 350 community leaders before him, he attacked them fiercely as cowards, who expected amenities and protection, but were not prepared to do anything to help the Government, as was their duty as citizens. There had been at least thirty incidents in the area in the last three months, sixteen guerrilla camps had been found, eight policemen had been killed and nine wounded, seven civilians murdered and two wounded. But only three people had come forward with any information.

'It does not amuse me', he declared, 'to punish innocent people, but many of you are not innocent. You have information which you are too cowardly to give.'

He then used his Emergency Powers to impose a collective punishment on the entire community in the form of a curfew for twenty-two hours out of every twenty-four and a reduction in the rice ration. At the same time, he gave the inhabitants a chance to give the authorities the information he was certain they had, and to give it without danger of insurgent reprisals, at least against individuals. Every householder was visited and given a sheet of paper, together with a personal letter from the High Commissioner and instructions to write down anything that he or anyone in the house knew about the Communists in the area. Twenty-four hours later, the papers were collected and inserted unseen into sealed boxes, which were then opened by the High Commissioner himself at Government House, in the presence of representatives from the villages.

This technique led to the arrest of some forty suspects and on 10th April the curfew was lifted. On 26th April, Long Pin, a noted guerrilla in that area, was killed, whereupon Templer visited the village again—this time to congratulate them.

The punishment meted out to Tanjong Malim had been severe, and undoubtedly many innocent individuals had suffered; but it had achieved valuable results. There was some criticism in England, that it was unwarranted and that this was what always happened when soldiers were given too free a hand; Cromwell's ghost stalked the lobbies of Whitehall for a while! In Malaya, however, the firm action was generally approved, and it certainly made the Malayan population begin to realize what was required of them.

By the end of 1952, there was a marked change in the situation, with a rise in confidence among the populace and the Security Forces, and a decrease in the number of incidents from 6,100 for 1951 to 4,700 in 1952; and the next year it was to drop to a mere 1,100.

During 1953, Templer made the Home Guard wholly responsible for the defence of their villages, a move which held risk, but which in fact paid off by increased morale and confidence, and by the freeing of fresh troops for offensive operations.

In the autumn of 1953, he introduced a dramatic policy of 'white areas', under which districts where the people had proved themselves actively opposed to the guerrillas were declared 'white' and freed from most of the irksome restrictions imposed on the remainder of the country. They would no longer suffer curfews, curtailed shopping hours, food control and prying patrols, but would be free and privileged.

The first spot chosen for this distinction was twenty-two square miles of the State of Malacca. On 3rd September 1953, General Templer declared the district to be a 'white area' and told the population that it was now up to them to 'keep the Communists out and see that they never come back', if they wanted to remain themselves free of restrictions. It worked, and over the next six years, the whole country was gradually to acquire this sought-after status.

As the guerrillas withdrew deeper into the jungle, they came to depend more and more on the 100,000 aborigines based there. These primitive, simple-minded people lived a life so divorced from reality that some of them thought that the Japanese were still in occupation of Malaya and they welcomed the Communists as resistance fighters against the invaders! The Communists soon brought the aborigines under control and forced them to provide food and shelter, and to help in many other ways. Clearly, this had to be stopped, and a plan was developed to free the aborigines from this domination.

First, patrols were sent into the isolated areas where the aborigines had their villages and patches of agriculture. Once there, the next step was to win their confidence sufficiently to be able to remain in the area. This was achieved, after which dropping zones were hacked out of the forest and supplies dropped; the clearings were then expanded sufficiently for helicopters to fly in with dismantled tractors and other equipment, from which the Sappers built airstrips and 'jungle forts'. These lonely outposts were finally occupied by police teams, who set out to win over the aborigines and so gradually gain control of their area and make it untenable to the guerrillas. It was an imaginative and successful move, which reached its climax with the formation of an Aborigine Fighting Regiment, which went into action in 1960 against the remnants of the Communists then lurking along the Malay-Thailand border.

In June 1954, General Templer left Malaya and his place was taken by his Deputy, Donald MacGillivray. Operational command of the Security Forces was then placed in the hands of a military Director of Operations, General Sir Geoffrey Bourne.

This was a reversion to the position before Templer arrived, but the situation had improved greatly since then, and the machine was running so much more smoothly that this division of responsibility was acceptable.

The Security Forces were now firmly on the offensive, and although the insurgents were still a considerable menace, they no longer presented any real threat to the Government. Their grip on the populace had weakened steadily, as the Army increased their pressure on the jungle gangs and as the police broke up the Branch organization in the populated areas. The increasing superiority of the Security Forces encouraged the populace and led to more and better information being provided; this in turn led to further successes against the insurgents, the flow of information increased, and the vicious circle had been broken.

Psychological warfare also contributed to the pressure against the insurgents by undermining their morale; one of the techniques was to employ insurgents who had either surrendered or been captured to persuade their former comrades to give themselves up. It was estimated that during September 1955, 50 million leaflets were dropped on the Communists, and that the total expenditure during 1955 to 1957 averaged about 50,000 leaflets per insurgent!

In September 1955, an amnesty was offered to the insurgents and to their supporters. It did not lead to any dramatic results, but it did improve the surrender rate markedly, and the numbers continued to increase as the insurgents became more and more disillusioned and disheartened.

November 1956 saw a good example of a successful large-scale operation against the guerrillas, when four battalions of 17 Gurkha Division cordoned off ten villages in South Johore simultaneously. Complete surprise was achieved and 278 suspects were arrested. Communist domination of the area was eliminated as a result, and information began to flow from then on.

Any hopes that the insurgents still held of winning the support of the people of Malaya faded completely when Independence was granted to the country on 31st August 1957; not a single

Communist or Communist supporter won a seat in the new Government.

But they still did not surrender, and it was to take a further three years of thankless, relentless campaigning to hunt down the surviving gangs widely scattered as they were in the vast areas of jungle and swamp. It shows no lack of appreciation of the Security Forces' achievements that the last five years of the campaign are dismissed in so few lines, but this phase produced few new lessons except for the great value of light aircraft, and helicopters in particular, in jungle warfare. In general, it was more a matter of continuing and improving already proven techniques, and keeping up the pressure.

By early 1960, it was calculated that only some 500 Communist insurgents remained in the Malayan Races Liberation Army, and they were lurking furtively along the Thailand border. On 31st July 1960, the Federal Government of Malaysia declared that the State of Emergency was at an end.

The campaign had been, as the C-in-C Far East Land Forces stated '. . . one of the longest periods the British Army had been on continuous active service since the Napoleonic Wars'.[1] A total of fifty infantry battalions from Britain and the Commonwealth had taken part, and the total casualties were 6,710 guerrillas killed, 1,500 captured and 2,700 surrendered; 1,865 members of the Security Forces and 2,473 civilians had been killed.

But Malaya had been saved from Communism, and a major cold war campaign had been won.

[1] *The Times*, 16th June 1959.

Chapter 7

MALAYA—GENERAL POINTS

The Enemy

The Malayan Emergency is a classic example of a Communist take-over bid, based on insurgency and guerrilla warfare. We shall doubtless see other attempts along similar lines, and it is, therefore, worth reviewing the enemy's aims and how they were thwarted.

The insurgents were 90 per cent Chinese, but they hoped to win over the Malayan half of the population by posing as a 'nationalist anti-British' movement. There was, however, no genuine anti-British feeling in Malaya, no strong nationalist feeling and no real local grievances to be exploited. The Communists failed to provide any cause which might arouse the sympathy of the Malays and they therefore never gained the willing support of the majority of the population.

The Communists lacked two other factors, which could have helped them. They had no political support for their military campaign, and so never managed to arouse world opinion to reinforce their military efforts, as Grivas was to do so successfully in Cyprus. They also lacked any friendly base outside Malaya, which they could use safely for reorganization and reinforcement. They had no border with China, as had the Viet Minh in Indo-China. Thailand to the north was hostile to Communism, and although the insurgents could lurk astride the frontier, they could not establish a firm base there. At sea, the Royal Navy controlled the Malayan coast. They were forced, therefore, to set up their bases in the Malayan jungle, which was unsatisfactory, as they had to be limited in scope and were always liable to attack.

The Communist plan of campaign was based on three phases. First, they would organize an insurgent movement by 'the masses' against the British Government and would adopt guerrilla tactics to tie down and demoralize the Police and the Army. Secondly, they would dominate or occupy certain specified districts, where they would create 'liberated areas' under Communist control. Finally, as the populace rallied to their cause, they would create a 'liberation army' and sweep the British from the country.

The insurgency never in fact got beyond Stage One, primarily because the Security Forces successfully broke up the large groups of guerrillas from the beginning, and so prevented them from setting up bases or dominating the vital areas around the populated districts. The insurgents could not, therefore, win over the masses as they had hoped, and they were in no position to set up 'liberated areas', except harmlessly in the jungle.

The Government then followed this up with their policy of isolating the guerrillas from the populace, and this finally defeated the revolt.

The Battle for Hearts and Minds

The Communists failed to win the minds of the people of Malaya as a whole, because of several errors of judgment. Firstly, they believed that the Malays were so discontented with British rule that they would willingly join the revolt against it. This was never the case, and was never likely to be, once Britain had made the promise in 1952 that the country would be granted independence in the near future. Secondly, they alienated the populace by their policy of sabotage, which caused unemployment and loss of income to the workers but did not win them the war. When the Communist leaders realized this, and issued the 1951 Directive ordering a policy of conciliation, it was too late. Finally, the only political solution they could offer to the Malays was a Communist State and this had little or no appeal.

The British Government in Malaya, on the other hand, early

appreciated the need to win the support of the populace, both in the short-term and as a long-term objective, and they achieved their aim. The Briggs Plan was a first step, when action was taken, first to protect and then to isolate the populace from the insurgents. This was followed in 1952 by the promise of independence, which was a trump card. It won over the Malays very effectively, though the Chinese needed more convincing, because they feared that they would be treated as 'second-class citizens' in an independent Malaya. They were convinced gradually, and once they turned against the insurgents, the revolt lost all hope of success.

The active participation of the populace in the campaign was encouraged by the establishment of a Home Guard, by recruiting for the police, and by constant exhortation. At first only the Malays responded, and the Police Force and the Home Guard were both almost entirely Malay. The Chinese sat on the fence until it was evident, firstly, that the insurgents were not likely to win, and, secondly, that support of the Government would be to Chinese interests.

A further feature was the official policy of improving the conditions of the people, both Chinese and Malay. 'Operation Service' was one example of it, and the social measures undertaken in the 'new villages' was another. Throughout the country, every effort was made to show that the Government had the interests of the people at heart, and this policy achieved results.

Alongside the battle to win over the waverers among the populace, there was the psychological warfare and propaganda campaign to win over the insurgents themselves, and this was also very successful. The insurgents lacked any propaganda or political support themselves, such as EOKA had in Cyprus, and this was a great weakness.

The Malayan Emergency illustrates well the importance of winning the hearts and minds of the people in the short and the long term, and how this can be achieved by the Government.

Civil-Military Co-operation

The Malayan Emergency was the first large-scale counter-insurgency campaign, and much pioneer work was required before the most effective machinery was worked out and put into effect. A good example of this was the problem of civil-military co-operation, particularly between the Army and the Police.

At first, the Army was 'in support of the Police', as if the situation were only a larger-than-usual 'internal security' operation. This soon proved unsatisfactory, and a system based on the triumvirate of 'civil, military and police' at all levels came into being as part of the Briggs Plan. This was developed into an extremely efficient machine, which became the model for the Kenya and Cyprus Emergencies.

This teamwork extended to Intelligence, where all sources and resources were co-ordinated under a single Chief of Intelligence, so that there was eventually very effective liaison, with the result that by 1954 the organization was good.

An important feature was that in the later stages the Malay and Chinese officials were included in much of the discussion on policies for the conduct of the campaign, and this helped to develop confidence between them and the Security Forces. Relations with the European population were close from the start, and the planters and miners played a decisive part by their stubborn resistance to terrorism and subversion.

The Malayan campaign eventually produced the highest standards of civil-military co-operation, and demonstrated that it is a decisive factor not only in defeating the insurgents, but also in establishing a political solution. The machinery did, however, have to be built up after the Emergency had been declared, and the twelve years' struggle might well have been shorter, or indeed never have got going at all, if the co-operation that was finally attained had existed in 1948.

THAILAND
PERLIS
Alor Star
KEDAH
Betong
Kota Bharu
South China Sea
PENANG
Grik
KELANTAN
TRENGGANU
Kuala Trengganu
PERAK
Ipoh
Cameron Highlands
Kuala Lipis
PAHANG
Kuantan
Fraser's Hill
Tanjong Malim
Batu Arang
SELANGOR
Kuala Selangor
Kuala Lumpur
Kajang
NEGRI SEMBILAN
Seremban
MALACCA
Malacca
JOHORE
Kluang
Johore Bahru
SINGAPORE
Straits of Malacca
N
International boundary
State boundary
Railway
Main road
Highlands (6000 ft. or over)
50
100 Miles
N.S.H.

Map I. MALAYA

The Emergency in Kenya
(1952–1956)

F

Chapter 8

KENYA—PHASE ONE (1945–52)

In February 1952, The Queen, who was then Her Royal Highness The Princess Elizabeth, was on a visit to Kenya and stayed at Sagona Lodge near Nyeri in the cool forests of the Aberdare Mountains. It was there that she received the news of the death of her father, King George VI, and left Kenya the same day.

Sagona Lodge was in Central Province[1] and both the Provincial and District Commissioners responsible for that area had for some time advised against the royal visit, because of the security risk from a secret organization called Mau Mau, which they believed was causing increasing trouble among the local Kikuyu tribe. But a security check revealed no threat, and the Government in Nairobi decided that the visit should take place as planned.

No incidents did occur, and in the same month the Governor, Sir Philip Mitchell, stated that the general political feeling in Kenya was better than he had known for many years.

Seven months later, a State of Emergency had been declared in the colony, Mau Mau terrorists dominated much of Central Province, and the Aberdares had been declared a Protected Area, in which anyone other than the Security Forces was liable to be shot on sight.

The origins of the Mau Mau revolt are many and complex and are fully analysed in the lengthy Corfield Report.[2] A detailed

[1] See map, p. 113.

[2] *Historical Survey of the Origins and Growth of Mau Mau* by F. D. Corfield (Command 1030 of May 1960).

study of this aspect of the Emergency is outside the scope of this book, but it is of value to look briefly at the situation in Kenya at the time when the revolt broke out, and to outline broadly the motives behind it.

Kenya in 1952 was a British Colony, ruled by a Governor on behalf of the Colonial Office in London. It was only fifty years since the country had been discovered by the first explorers, who had found it occupied by primitive, nomadic tribes, constantly at war with each other, and possessed of no national concept or loyalty outside their own tribes. Although in 1952 the Africans outnumbered the whites by almost 100 to 1,[1] there was little or no thought then that they were as yet in any position to take over any control of the country, and it was assumed by the white population that British rule would continue for many decades yet.

The Administration under the Governor was headed by a Legislative Council, which was in effect the Parliament, and consisted entirely of nominated officials; the European population was represented by Elected Members, but even they sat on the Opposition bench, from where they were able to express their views but had no executive power. The Africans had no elected representatives at all, but were looked after by an official, known as the Chief Native Commissioner. From the Legislative Council the Governor formed his Executive Council, which was his Cabinet.

For administrative purposes the colony, which covers 225,000 square miles (twice the area of England and Wales) was divided into seven Provinces, which in turn were subdivided into Districts. Each Province was administered by a Provincial Commissioner served by District Commissioners and District Officers, who exercised an autocratic but benevolent control over all aspects of life in their own areas.

The police were organized on a similar hierarchy, under the Governor. The officers were European, with African constables, while in the tribal Reserves, a special Tribal Police Force was

[1] Population figures 1952: African, 5,500,000; Asian, 165,000; European, 55,000.

created consisting of Africans of that tribe. There was an efficient, but very limited, Special Branch, and a large Police Reserve composed of the European settlers.

The Army in Kenya consisted of three battalions of the King's African Rifles,[1] again commanded by European officers with African N.C.O's and men. They were well disciplined and experienced troops, many of whom had fought in the Second World War, and their loyalty was unquestioned. The soldiers came almost entirely from tribes other than the Kikuyu, not as a matter of Government policy, but because the Kikuyu temperament preferred the pen to the sword.

There was also the Kenya Regiment, which was an officer-training battalion consisting only of European volunteers.

Some two-thirds of the Europeans in Kenya were farmers in the famous White Highlands;[2] this area comprised about one-third only of the best land in the country but it produced four-fifths of the colony's agricultural exports. The settlers had built up their farms from virgin bush against tremendous difficulties; they farmed well and looked after their African labour well. They also guarded their rights jealously, and no Africans were allowed to buy land in the White Highlands.

The Africans were just beginning at this time to think of their rights, but only one tribe was showing any real political consciousness. This was the Kikuyu, who were the most sophisticated and educated of the tribes.

The Kikuyu Reserve bordered on the White Highlands in many areas, and envious African eyes were inevitably cast on the rich crops and sleek cattle of the Europeans. This envy was aggravated, not only by the special significance attached by the Kikuyu to the possession of land, but also by the existence in their minds of a long-standing grievance to the effect that the Europeans had taken their lands from them. This allegation had been meticulously investigated by the Government and proved unfounded, but it had persisted nevertheless in the Kikuyu mind as an ember

[1] 5th Bn at Nanyuki, 7th Bn at Nairobi and 23rd Bn at Archer's Post.

[2] The White Highlands were fertile plateaux of 6,000 to 9,000 feet, mostly located in Central and Rift Valley Provinces.

of resentment, and was to be deliberately fanned by the Mau Mau leaders into the flame of revolt.[1]

The Kikuyu Reserve also bordered on Nairobi, and this led the Kikuyu to become urbanized and de-tribalized more rapidly than any other tribe. It led them furthermore to become envious of European privilege, resentful of European power and position, and bitter over the racial discrimination which denied them access to society, to land and to the better Government jobs.

These emotions were common to many African races, which had had their eyes opened by their experiences in the Second World War. The 'wind of change' was just beginning to blow, and its force depended on the leaders that it found to harness it. In Kenya, nationalist ideas were most active in the Kikuyu tribe, but even there, they were largely quiescent, until they were stirred up by the arrival in 1946 of Jomo Kenyatta.

His activities were purely political and centred round a revival of the main African political party, the Kenya African Union. But behind this legal 'front' the illegal and subversive Mau Mau organization was developing, with the aim of gaining by violence what might not be obtainable by peaceful means.

The Mau Mau movement was not a national movement, because it was confined almost entirely to the one million Kikuyu, who represented only one-fifth of the African population of Kenya. There is no evidence that it was Communist inspired or even received material Communist support. It was a deliberate programme of insurgency, designed to enable its supporters to dominate, first, the whole Kikuyu tribe and then the other tribes in Kenya. The ultimate aim was to eliminate the whites in the colony and establish an independent, Kikuyu-controlled Kenya.

Reports of Mau Mau activities and the administering of oaths began to appear in 1950, and in April of that year, nineteen Africans were tried and convicted for this offence, (though they were later acquitted on a legal technicality).

From the start, Mau Mau used 'oathing' as one of their main weapons, firstly, to control their own members, and secondly, to intimidate those who were reluctant to join the movement.

[1] *Corfield Report*, Ch. II.

The taking of oaths had long been an accepted part of Kikuyu custom, and the tribe believed firmly in the killing power of an oath—a belief which the Mau Mau effectively strengthened by some carefully chosen murders. The oaths served not only as an ideal method of imposing discipline, but also ensured that Mau Mau security was of the highest standards, for the first oath taken by all recruits was:

> 'If I ever reveal the secrets of this
> organization, may this oath kill me.'[1]

Similar oaths ensured loyalty and obedience to the Mau Mau leaders and a promise not to co-operate in any way with the Government; one oath bound the victim to pay his gang subscription regularly, which must have eased the task of the Gang Treasurer considerably!

Initially, the oaths were purely negative, requiring secrecy, loyalty and obedience, but gradually they began to demand action:

> 'If I fail to steal anything I can from a European may this oath kill me,' and 'If I am called upon to do so, I will kill a Kikuyu who is against Mau Mau, even if it be my mother, or my father, or brother or sister or wife or child.'[2]

The forcible administration of oaths was a very effective form of intimidation; the victim was deeply reluctant to break his word, largely because those who had taken these oaths felt themselves to be outcasts from the tribe and therefore more heavily committed to Mau Mau. This was part of a deliberate policy on the part of the leaders of the revolt, who sought to undermine and break the authority of the tribal elders by deliberately deviating from the established tribal customs, and substituting the increasingly vile and demoralizing oaths, which became such a feature of the later stages of the revolt.

In 1951, warning reports of Mau Mau subversion increased, but were largely discounted by the Central Government; 1952

[1] *Corfield Report*, Ch. VI.
[2] Ibid.

saw the royal visit in February, and then from June until September, the colony was unfortunately without a proper Head, as the outgoing Governor was taking his terminal leave and his successor could not take over until this was completed. During this interregnum, the Central Government was constantly warned of an 'increasing disregard of law and order', and of 'a general revolt among the Kikuyu people'.[1]

But little effective action was taken, and on 19th August, the Member for Law and Order stated: 'I categorically deny that there is a state of emergency.'[1]

Then on 28th September, the new Governor, Sir Evelyn Baring, arrived in Kenya. He was sworn in the next day and immediately set off on a tour of the colony. Nine days later, he emphatically recommended to London that a State of Emergency should be declared in Kenya as soon as possible.

At 8 p.m. on 20th October 1952, the Governor signed the relevant documents, but no public announcement was made. The three battalions of the King's African Rifles already in Kenya were, however, alerted and action was taken to reinforce them by a further two battalions from Tanganyika, Mauritius and Uganda.[2] At 7.30 p.m. the first of twelve RAF Hastings aircraft landed at Nairobi, bringing the leading elements of the Lancashire Fusiliers from the Canal Zone. Then between midnight and dawn, the police rounded up ninety-nine Mau Mau suspects, including Jomo Kenyatta, and quietly arrested them.

The next day, 21st October 1952, the Governor made a broadcast to the colony announcing a State of Emergency.

[1] *Corfield Report*, Ch. V.

[2] 4th Bn ex Uganda, 6th Bn (less two companies) ex Tanganyika and two companies of 26th Bn ex Mauritius.

Chapter 9

KENYA—PHASE TWO (1952–54)

A sense of anti-climax followed the declaration of the Emergency. The round-up of suspected Mau Mau seemed to have thrown the insurgents temporarily off balance, and there was no large-scale, planned revolt. This was perhaps just as well, since the responsibility for the control of security operations still rested with the Attorney General, who was at the same time the Member for Law and Order (equivalent to the British Home Secretary); there was no such post then as Minister of Defence or Internal Security.

However, even if he was somewhat overwhelmed on the 'order' side of his dual appointment, he coped effectively with the 'law' aspect, and passed some Emergency Regulations which dealt with the control of both the 'active' and the 'passive' wings of Mau Mau. Other measures initiated during this welcome lull were the embodiment of the Kenya Regiment and the calling up of the Kenya Police Reserve.

The European settlers undertook, as might be expected, to look after themselves; with limited help from the Army, they soon set up an excellent system of local defence for their widely scattered farms, and thus freed troops for more active duties. Farm Guards were organized from among the loyal African labour, stock was protected as far as possible, and a system of rockets and other signals brought help swiftly to the scene of any attack. The greatest credit is due to these settlers, and even more to their womenfolk, who refused at any time to be daunted by the constant threat of murder, arson and even torture.

The threat was very real. It was impossible to tell at this stage which Kikuyu were loyal to the Government and which were Mau Mau; some had joined the rebels willingly, but many more followed them because of intimidation in the form of threats to themselves and their families—threats which they knew were not idle ones.

The insurgency involved the whole of Central Province and the eastern part of the Rift Valley; this included the two vast forested areas of the Aberdare Mountains and Mount Kenya, both of which provided ideal bases for guerrilla activities, and were used as such by the 'active wing' of Mau Mau. There was little terrorism in the towns, and virtually none there against Europeans; another feature of the campaign was the failure of the Mau Mau ever to initiate any sabotage against the Security Forces. Counter-insurgency operations would have been very much more difficult if there had been determined attacks against the railways, roads, power installations and communications generally.

The insurgency was organized from Nairobi, which was the base of the Central Committee; this body was responsible for overall policy and controlled the campaign in the Reserves and the forest areas through District Committees.[1] But the initial wave of arrests on the declaration of the Emergency had dislocated their plans and it was a while before the insurgents were again in a position to start active operations.

The military forces of Mau Mau called themselves the 'Land Freedom Army', and they consisted of a guerrilla force which numbered between 12,000 and 15,000 at its peak. It was organized, at least on paper, into a number of formations imaginatively described as 'Armies' commanded by 'Generals', but which were in fact gangs of 50 to 100 Mau Mau, which were more accurately termed 'platoons'. Only about 12 per cent of them were armed with precision weapons, and their deadliest tool was the razor-sharp matchet, called a 'panga'. They had received virtually no military training, but possessed a hunter's instincts and skill, as well as the cunning of a tribe whose home had always been in or near the forest. As a result, one of the main problems of the Security Forces throughout the campaign was how to make con-

[1] See Appendix C, columns (*e*) and (*f*).

tact with these guerrillas in an element which was so alien to the average soldier and so natural and advantageous to the Mau Mau.

The 'active wing' of the insurgency was supported by a 'passive wing' which was thought to number about 30,000; they were based in the Kikuyu Reserves and were organized by the District Committees to carry out the usual tasks of providing food, money, information and recruits for the guerrillas.

The Security Forces were not prepared at first for co-ordinated action against the insurgents. The Army acted initially in support of the police and was dispersed throughout the colony on defensive tasks—so scattered indeed that one Battalion Commander found that he had to travel 400 miles to visit all his companies. Headquarters East Africa was not geared to active operations and there was as yet no effective civil-military command structure.

Similarly the Civil Administration lacked any effective organization to co-ordinate and control the Emergency. The senior policy-making body was the Sitrep Committee which was more accustomed to the routine discussion of routine reports; it was to be reorganized before very long as the Emergency Committee and it did its best to cope with the situation; but it could not be the executive body that was required.

Any active measures against the revolt were inhibited from the start by lack of information. There was no effective joint intelligence organization in the colony before the Emergency and the Special Branch was totally inadequate for the task now facing it. This was a period of frustration and improvization for the Security Forces, which was to last until February 1953.

One step of great long-term significance which was made at this time was the decision to form a Kikuyu Home Guard. It was a brave and imaginative move on the part of the Administration to set up, and later to arm, members of the tribe which had given birth to Mau Mau, and 90 per cent of which had taken some form of Mau Mau oath. But it was proved to be a right decision. Within a few months, the Home Guard numbered 10,000, and then rose to 20,000; they fought resolutely against their own tribe, first with spears and pangas, and only later with shotguns and rifles. Many of them were killed; but together with the Kikuyu Tribal Police, they had by the end of the Emergency killed no less than 4,686 Mau Mau, which amounted to 42 per cent of the total bag.

The real significance of the Home Guard was that they prevented the Mau Mau revolt from spreading, and also limited the insurgent attempts to intimidate the populace. As a result of their resolute defiance from the start, the rebellion became a civil war within one tribe instead of being a nationalist movement. For this reason it is no exaggeration to say that the Home Guard was a major factor in the final defeat of Mau Mau, and this is a lesson not to be forgotten in any counter-insurgency campaign. A further valuable step was that a registration of the male population had been completed and identity cards issued to all male Kikuyu in December 1952.

January 1953 saw the murder of two settlers at Ol Kalou, and the shooting of an African Chief while he lay in a hospital bed; then a complete European family was murdered and savagely mutilated near Kinangop. By mid-February, Mau Mau had, since the start of the Emergency, killed nine Europeans, three Asians and 177 Kikuyu; the only notable success against them had been scored by two intrepid women settlers, Mrs. Hesselberger and Mrs. Raynes Simpson, who killed three of a gang which attacked them as they sat at supper in their isolated farm.[1]

This outbreak of atrocities by Mau Mau in the New Year, and the apparent inability of the Government to deal energetically or effectively with it, led to a growing sense of dissatisfaction throughout the colony. The particularly savage and senseless killing at Kinangop sparked off an explosion of fury, which culminated in an angry demonstration at Government House by some 1,500 Europeans, who were determined to express to the Governor himself their demand for a more vigorous prosecution of the Emergency.

This was the nadir of the campaign in Kenya. Even as the angry settlers shouted for the Governor to appear and speak to them, the measures which they demanded were already in train, and would be implemented within a few weeks. The Sitrep Committee was reorganized, as already mentioned, into the Emergency Com-

[1] Both were subsequently awarded the M.B.E. for their gallantry.

mittee. The Attorney General was relieved of his responsibility for internal security, which passed to the Chief Secretary (the equivalent of the Prime Minister, under the Governor). Most important of all, Major-General W. R. N. Hinde was appointed as Military Adviser to the Governor, a title changed within a few weeks to Director of Operations.

From that moment on, the situation improved. The bitterness of the up-country settlers subsided, and their morale improved; the Security Forces too took heart and began to think of getting to grips with the enemy. At the same time, the command structure was reorganized, and the essential triumvirate of 'civil, military and police' was developed and strengthened at all levels. Everywhere the nuts and bolts of the Emergency machine were tightened where necessary, the engine was overhauled, topped up and set in motion; it still creaked a bit, but at least it was moving.

Then came 20th March 1953—a black day for Kenya. At 9.30 p.m., 1,000 Mau Mau descended on the loyalist Kikuyu village of Lari, some twenty-five miles north-west of Nairobi, and massacred eighty-four of its inhabitants, two-thirds of them women and children; they also horribly mutilated thirty-one others, burned 200 huts and maimed 1,000 cattle.

At the same hour, only thirty miles away, another Mau Mau gang raided Naivasha Police Station, killing two policemen, stealing forty-seven weapons and releasing 173 prisoners.

These two humbling setbacks to the Government had a profound effect on the future of the Emergency. The Naivasha incident put the Kenya police on their mettle as had nothing else, and it led to the hastening of some long-overdue improvements in their efficiency. Lari shocked Kenya and the world, and it also demonstrated beyond all doubt that Mau Mau was directed against its fellow Africans rather than against a Colonialist power.

An even more significant outcome of the Lari massacre was that many Kikuyu who had till now wavered and given passive support to the revolt were so sickened and shocked by the savagery perpetrated there that they turned away from Mau Mau. From that moment, recruiting to the Home Guard increased and the decision was taken to arm them; from that moment, information began to flow in to the Security Forces, and from that moment, Mau Mau may be said to have lost the war, though there was a long road still ahead.

A further major change occurred on the 7th June 1953, when General Sir George Erskine was appointed Commander-in-Chief. He also became Director of Operations, and Major-General Hinde was made Deputy Director of Operations; Major-General G. D. G. Heyman arrived as Chief of Staff. At the same time, the Command ceased to be subordinate to Headquarters Middle East Command in Cairo and became an independent Command, responsible direct to the War Office.

The C-in-C was given 'full powers of command over all Colonial, Auxiliary, Police and Security Forces'. He did not, however, as in Malaya and Cyprus, become responsible for civil administration as well as military operations; he remained 'in support of the civil power' throughout the Emergency, but with the Police Force now under his operational command. The Governor continued in overall control of the Administration, because the Emergency affected only one-sixteenth of the whole country and one-fifth of the population. He therefore administered the colony as a whole, while the C-in-C was responsible for security operations against the insurgents. In this instance, the divided responsibilities worked satisfactorily.

Reinforcements for the Security Forces had arrived in April in the form of 39 Infantry Brigade, and the troops at the disposal of the C-in-C now amounted to:

39 Infantry Brigade	Two British battalions
70 Infantry Brigade	Six Battalions of King's African Rifles (KAR)
Additional Troops	(*a*) One British battalion
	(*b*) One East African Independent Armoured Car Squadron (EAAC)
	(*c*) 156 HAA Battery, East African Artillery
	(*d*) The Kenya Regiment

The C-in-C appreciated on arrival that it was time that the Army turned to the offensive, and he therefore prepared a plan based on this aim. It was the first time that the Security Forces had had a clearcut policy, let alone a chance to seize the initiative, and it was a welcome change after the usual defensive opening period. It was not a powerful offensive that was mounted, but it was a gesture based on the old principle: 'Don't just stand there. Do something!'

The plan was that 70 Brigade with its African battalions should continue to operate in the Kikuyu Reserve. From the remainder of his command, the C-in-C managed to raise three rather meagre 'striking' forces:

(1) 39 Infantry Brigade and the Kenya Regiment.
(2) The EAAC Squadron and elements of 156 HAA Battery.
(3) The RAF.

39th Infantry Brigade were tried for a short time on operations in the Kikuyu Reserve, but it was found that this was a task for which the African units were better suited. The British battalions were therefore employed in the forest areas of the Aberdares and Mount Kenya, which were 'Prohibited Areas', where anyone seen other than the Security Forces could be shot on sight. The search for Mau Mau guerrillas in this 1,500 square miles of thick forest and undergrowth was more forlorn than looking for the proverbial needle in the proverbial haystack, but it provided useful training and experience, and at least it was offensive action, which was good for morale.

The mobile column, consisting of the EAAC Squadron and the gunners, was used in the open country, while the RAF made their contribution by bombing the forests. It is not thought that these air attacks killed many guerrillas and the policy was condemned by many as a heavy-handed waste of effort. But later interrogations did indicate that it had a demoralizing effect on the Mau Mau and also harassed them by making them move around in the forest. General Erskine maintained firmly that it was worth a further brigade of troops to him. Unlike bombing in a populated area, it did no harm to property or to civilians, and it was therefore, to my mind, justified, if it assisted the Security Forces in their task, which it must have done to some degree.

Operations were still very much a matter of trial and error at this stage, and it was discovered, among other things, that it was impossible to penetrate into the forest without the guerrillas knowing it and disappearing before contact could be made. It was essential therefore to have bases well inside the forest, and this was achieved by cutting rides to a depth of 7,000 yards, at the end of which battalion camps were established as patrolling bases. The Public Works Department undertook this work at first, but their other commitments were so heavy that it was found necessary to send 39 Field Engineer Regiment out from England to assist them.

The 'striking forces' were very thin on the ground at first, but by August the KAR Battalions could be freed from the Reserve, and in October, when a new formation was made available in the shape of 49 Infantry Brigade, the offensive gathered momentum.

By the end of 1953, the 'active' wing of Mau Mau, now estimated at 12–15,000, had lost around 3,000 killed and 1,000 captured. The 'passive' wing too had suffered, and more than 150,000 Kikuyu, with a proportion of their Embu and Meru neighbours, had been arrested; of these, 64,000 were brought to trial. On the strength of these encouraging figures, the C-in-C felt able to claim that a start had been made on restoring confidence in the Government's ability to restore law and order.

By now, the Command was at its maximum strength of five British and six African battalions, together with the Kenya Regiment, auxiliary units and the RAF, a total strength of about 10,000, supported by 21,000 Police and about the same number of Home Guard. With these forces at his disposal, the C-in-C was in a position to work out a long-term plan for 1954.

The position at the beginning of that year was that the Mau Mau were far from being crushed, but they had no chance now to dominate the whole country or even the whole of their own tribe. They had killed or wounded almost 1,000 fellow Kikuyu, as against 21 Europeans, but they had failed to win the tribe as a whole over to their cause. And it was now too late, for the initiative lay at long last with the Security Forces, and they were to take full advantage of it.

In March 1954, an important change was made in the command structure for the control of the Emergency. The unwieldy Emergency Committee was changed to a small, executive War Cabinet consisting of the Governor, the Deputy Governor, the C-in-C and Michael Blundell who was the Chairman of the European community.

The War Cabinet met twice weekly and was served by an Executive Officer, who ensured that action was taken on Cabinet decisions, and by a Secretariat, consisting of both civil and military officers. An Emergency Joint Staff of civil, military and police officers was also established to co-ordinate planning on behalf of the War Cabinet. Below them, the civil-military-police triumvirate continued to function effectively at Provincial and District Emergency Committees, which planned security operations; command and executive control of units rested as always with the military and police commanders concerned. This organization worked well, and lasted until the end of the active period of the Emergency.

By early in 1954, it had become apparent that the centre of the Mau Mau organization lay, in fact, in Nairobi. The capital city had a population of 200,000, of whom some 65,000 were Kikuyu; about 30,000 of these had moved into the city since the start of the Emergency, and all of them were sympathetic to the revolt. Nairobi undoubtedly formed the base from which the gangs in the forest were supplied with food, funds, weapons and, of course, recruits. The Mau Mau Central Committee with the military Headquarters was also there and it was the brains behind the whole revolt. But this activity was all 'underground' and extremely difficult to eliminate.

Lawlessness was rife and blatant, and the extent of the Mau Mau control over the city was highlighted by an incident when a Mau Mau 'trial' was held in broad daylight on two loyal Kikuyu nurses who had just been abducted from a nearby teashop for the alleged offence of being 'collaborationists'. They were duly condemned to death on the spot by the 'divisional court judge' and only escaped by attracting the attention of the police as they were being dragged down the street for summary execution.[1]

It was decided that a vast 'cordon and search' operation to clean up Nairobi must be the primary operational task for 1954. It was a major undertaking and there were preparations to be made that could not be concealed. A reception camp for 10,000 was completed at Langata, five miles outside Nairobi, while detention camps for 10,000, later increased to 24,000, were prepared at the old wartime base of Mackinnon Road, 250 miles towards Mombasa. It was realistically accepted that it would not be possible to maintain complete secrecy regarding the operation, but surprise was achieved as to the actual date, which was 24th April 1954.

The plan for 'Operation Anvil', as it was called, was in accordance with normal procedures, the Army cordoning off the area, while the police carried out the actual searches. Five battalions and part of the Kenya Regiment were used to form a thin but effective screen, while the city was thoroughly combed by the police, one sector at a time. As each area was cleared, the Army took control of it, until the whole city had been dealt with in this way.

The problem was how to sort out the innocent from the guilty. It was not possible to remove and screen the entire community, and the procedure finally adopted was to screen all those who could show no real and useful need for being in Nairobi, detaining those who showed signs of needing more thorough checking. In the first major phase of the operation which lasted two weeks, about 30,000 were screened, and of these, just over half were detained; in addition, some 2,500 of their dependents were evacuated to the Reserve.

A second two-week phase then began on a smaller scale to 'tidy up' and, at the same time, forgery-proof identity cards were issued to those remaining in Nairobi, as an aid to future control of movement in the city.

'Operation Anvil' has been criticized as being unduly ruthless, indiscriminate and clumsy, but it was undoubtedly a success. It

[1] White Paper Cmd 9081, *Report by Parliamentary Delegation to Kenya*, January 1954.

achieved three major results quickly and effectively. First, it cut off the guerrilla gangs from their main source of supply and from their higher command, which was itself severely disrupted by the removal of many of its senior members. Secondly, it smashed the grip of Mau Mau on the city and they never regained it; crime dropped dramatically, terrorism was reduced and, from then on, the Security Forces were in firm control. Finally, and perhaps most important of all, the operation provided a vast amount of information about Mau Mau. After 'Anvil' a complete picture could be built up of the organization not only in Nairobi but also in the Reserve and in the forests. This was to be used from then onwards to enable the Security Forces to operate against specific targets, instead of indiscriminately as had been necessary to date. Moreover, now that Mau Mau had suffered a major defeat on a large scale, the co-operation of the local population with the Security Forces improved and more and more information began to be supplied voluntarily.

'Anvil' was a turning point in the Emergency and it may even be said to have broken the back of Mau Mau, who were thereafter increasingly demoralized and on the defensive, although it was to take another two and a quarter years before it was possible to declare that they no longer constituted a threat to security.

Firm action was taken to follow up the success achieved by 'Anvil'. Firstly, pressure was maintained on Nairobi by stationing a battalion there, and by carrying out a series of surprise swoops, based on the much improved intelligence now available. Secondly, the frayed link between Nairobi and all other Mau Mau areas was broken permanently and completely by a further series of 'screening' operations against the Reserve. It was now possible to work outwards from Nairobi as a secure base and to clear the Reserve, thus driving the gangs into the forest areas, and at the same time cutting them off from both Nairobi and the Reserve. As each area of the Reserve was cleared, it was handed over as far as possible to the Civil Administration and the Home Guard, so as to free troops for active operations in the forest.

This policy was reinforced by stepping up two other measures already in operation, 'food denial' and 'resettlement'. Food denial followed closely the methods used in Malaya and was equally effective; as supplies became short in the forest and no help came from the Reserve, the gangs were forced to emerge in order to forage for their essential needs. By forcing them to move as far as possible from the forest, it became easier to ambush them; and to make things even more difficult, a great ditch, fifty miles long, wired and booby-trapped, was dug along the forest edge and guarded by police posts every half-mile. This measure did involve some hardship to the European farmers near the forest, because they were forbidden to grow food crops within three miles of the forest edge and had also to keep their cattle enclosed at night, which reduced the milk yield. Some farmers calculated their losses as high as 25 per cent; but once the necessity for these restrictions had been explained, they co-operated loyally, and the measure did much to help to control the gangs.

Another security measure, based on experience in Malaya, was the resettlement of scattered Kikuyu families into 'new villages' where they could be better controlled and guarded; this had been begun early in the Emergency as a primarily defensive measure, although it had been opposed by many on the grounds that it was directly against Kikuyu tribal custom. Initially it was accepted only reluctantly by those who were required to move; but once its advantages, such as improved shops and schools, as well as increased protection, came to be realized, the project developed rapidly. By the end of 1954, about one million Kikuyu had been resettled, thus greatly assisting both the Security Forces and the Home Guard in their tasks.

The year 1954 therefore saw the decisive blow struck against Mau Mau in the form of Operation 'Anvil', followed up by action which drove the remnants of Mau Mau into hiding in the forests, where they would be hunted down and steadily destroyed. The Security Forces were now unquestionably on top, and the end of 1954 can therefore be regarded as the end of Phase Two.

Chapter 10

KENYA—PHASE THREE (1955–60)

Phase Three began with two large-scale operations against the Mau Mau in the forest. The first was aptly named 'Hammer' and consisted of a vast sweep by a division's worth of troops through the Aberdares. It lasted a fortnight during January 1955, and resulted in only 161 Mau Mau killed, wounded or captured.

It was followed by 'First Flute', which was on much the same scale; it lasted almost the whole of February and March, and was directed this time against the forests round Mount Kenya. A new technique was also tried out; instead of one vast sweeping operation, each unit was given its own area of forest to dominate; this was an improvement and the bag this time was 277 Mau Mau. Considering the very large numbers of troops involved, the results of both these operations were disappointing; admittedly these tactics were gradually wearing down the morale and the fighting efficiency of the gangs, but it was a clumsy and costly process—it was estimated that every dead Mau Mau cost £10,000![1]

Meanwhile Special Branch had developed their own technique, which was not only very much cheaper, but was also far better suited to the changed circumstances, where the enemy now consisted of small, fugitive gangs, hidden deep in the forest. It had been found that the loyalty of most Mau Mau tended to be

[1] This seemed a high price at the time, but it has recently been estimated that every dead Vietcong in Vietnam costs the United States $373,000 (£133,200) (*Weekend Telegraph*, 3rd June 1966).

directed towards an individual leader rather than to a cause; therefore, they were often perfectly willing when captured to 'turn their coat' and to operate under European leadership against their former comrades in arms.

Full advantage was taken of this, and former Mau Mau guerrillas were therefore being increasingly used as guides to European patrols or even to form 'pseudo-gangs' of Mau Mau, which penetrated deep into the forest, luring the surviving guerrillas into traps and ambushes.[1] The results were highly encouraging, and there seemed no doubt that this was the ideal technique for the closing stages of the campaign, when the gangs, like coveys of October grouse, were becoming smaller, more scattered and more cunning.

The proposal to use ex-Mau Mau was approved initially by General Erskine, and it was then developed considerably by his successor, Lieutenant-General Sir Gerald Lathbury, who took over as C-in-C on 2nd May 1955. The new C-in-C quickly appreciated that the time for large-scale operations was past, and that the surviving guerrillas would only be rounded up by cunning rather than by massive manœuvres. He therefore gave every encouragement to the use of 'pseudo-gangs', and immediately authorized the Commissioner of Police to raise five Special Force Teams, as they were now called.

Until then, the 'pseudo-gangs' had always been led by Europeans, but General Lathbury agreed to back Superintendant Ian Henderson when he proposed sending them into the forest armed and alone. It was a grave risk and could have had serious results if they had decided to desert with their weapons—and there were many who were convinced that they would. All did go according to plan, however, and this technique led to the elimination of many of the Mau Mau leaders, who might otherwise never have been located. Indeed, these operations may well have shortened the Emergency by several years.

All efforts were now concentrated against the forest. Selected and limited sweeps kept the gangs on the move, while battalions, now well trained in forest warfare, divided their efforts between

[1] *Gangs and Counter-Gangs* by F. E. Kitson.

patrols within the forest and ambushes along the edge. These tactics, together with the continued use of food denial and Farm Guards, led to the gangs becoming increasingly broken up and disorganized.

Towards the end of the year, a new technique was devised called a 'population sweep', where the District Officers encouraged all available Africans in an area to report, equipped with pangas, to the forest edge; 70,000 appeared on one occasion at Meru and 45,000 turned up elsewhere later. They were mostly women, and having lined up, almost shoulder to shoulder, they literally hacked their way, singing and laughing, through the forest; it was a huge joke to all participants, except any hapless guerrilla who might be flushed, for if he was not shot by the supporting police, he would literally be cut into very small bits by a pack of flailing pangas. The local population was now no longer 'passive' and the end of the revolt was near.

These tactics produced a guerrilla elimination rate of some 300 a month. By the end of 1955, the total Mau Mau still active was estimated at 2,000, compared to a peak figure of about 15,000. In September 1956, the enemy strength was down to 500 desperate, hunted individuals, who were no longer an organized force. Next month, the dramatic nine-month manhunt for Dedan Kimathi, the self-styled Mau Mau Commander-in-Chief, ended with his capture,[1] and with that the Mau Mau were to all intents and purposes eliminated as a threat to security.

The military forces had been steadily reduced in numbers ever since August 1955, and in May 1956, operational control was handed over to the police. Six months later, the Army was withdrawn from active operations.

The Mau Mau guerrillas had been decisively defeated on the field of battle, but the insurgency had not been a vain effort; the Kikuyu resort to violence in 1952 and the ever-present possibility of a recurrence thereafter greatly strengthened the hand of the nationalists during the political struggle for power which continued from 1956 onwards.

[1] *The Hunt for Kimathi* by Ian Henderson.

The final cost of the campaign was over £55 million, and the total casualties by the end of 1956 were:[1]

	Killed	*Wounded*	*Captured*	*Surrendered*	*Total*
A. *Security Forces*					
European	63	102	—	—	165
Asian	3	12	—	—	15
African	524	465	—	—	989
Total	590	579	—	—	1,169
B. *Civilians*					
European	32	26	—	—	58
Asian	26	36	—	—	62
African	1,817	910	—	—	2,727
Total	1,875	972	—	—	2,847
C. Total of A and B	2,465	1,551	—	—	4,016
D. *Mau Mau*	11,503	?	2,585	2,714	16,802

Although military operations had ceased by the end of 1956 the State of Emergency was not declared at an end until 12th January 1960—only three years before Kenya was to become an independent state, led by Jomo Kenyatta, whose arrest in 1952 had signalled the declaration of the Emergency. Phase Three of the insurgency in Kenya really ended on Uhuru Day in 1963, when Jomo Kenyatta was acknowledged as the head of an independent, Kikuyu-dominated Kenya.

[1] Hansard, 29th June 1959. These figures do not include detainees.

Chapter 11

KENYA—GENERAL POINTS

Surrender Offers

In the Kenya Emergency, the Mau Mau leaders did not make any truce offer on their own initiative, but the Government offered them surrender terms on three occasions.

The first offer was made in August 1953, less than a year after the start of the Emergency. It was designed to tempt away from Mau Mau those who had been forced into it against their will. The terms were that those who surrendered would not suffer the death penalty then in force for membership of Mau Mau. The appeal led to no immediate surrender on a large scale, but it did induce 815 guerrillas to surrender by the end of 1954.

In April 1954, the guerrilla leader, Waruhiu Itote, otherwise known as General China, was captured near Mount Kenya, a notable success, as only eight major Mau Mau leaders had been accounted for to date. Under the skilled and infinitely patient interrogation by Superintendent Ian Henderson, General China was persuaded to reveal the complete Order of Battle of the guerrilla forces in the Mount Kenya area. He then went on to suggest that he might be able to arrange a mass surrender of the Mount Kenya gangs. After much discussion, his proposal was accepted, and it was agreed that the reward for success was to be the remission of the death sentence that hung over him; the operation was appropriately named 'Wedgwood', in honour of its chief actor.

Secret negotiations were then begun and many meetings were arranged in the forest, attended among others by the Chief Native

Commissioner, representing the Governor, and the Chief of Staff, representing the Commander-in-Chief. The move was approved by the loyalist Kikuyu, but was bitterly opposed by most European settlers, who saw no virtue in coming to any terms with the Mau Mau.

After lengthy discussion, it finally seemed that several guerrilla leaders were prepared to surrender, together with their gangs, and the Government agreed that they would not be executed if they did. It was arranged that the surrender date would be 10th April, and that a 'cease-fire' would be put into effect between 30th March and 10th April in the Prohibited Area of the forest, but not in the Reserve adjoining the forest. Then on 7th April came one of those twists of fate; a large Mau Mau gang was reported at Gathuini, which was two miles outside the Prohibited Area. It was therefore considered by the Security Forces to be a legitimate target and was heavily attacked; but it was immediately alleged by the guerrilla leaders that this had been a gang trying to surrender, and that their fate proved that the Government had been laying a deceitful trap from the first.

As a result of this unfortunate incident, a mass surrender never materialized, and, to rub salt into the wound, two other Mau Mau leaders, who had been helping General China with the negotiations, took advantage of their position and escaped. General China himself stayed and was pardoned for his efforts.

When this all came to light, there was an outburst of indignant criticism from the public that three Mau Mau leaders had been allowed to escape with their lives, with nothing to show for it. The gains had, however, been far from negligible in the way of valuable intelligence, which was now to lead to some highly successful operations—and a mass surrender had been very close.

The final series of surrender negotiations began in February 1955 and dragged on until May. At one moment, they seemed near to success, but Dedan Kimathi got wind of it and promptly threatened to kill the negotiators on both sides; whereupon the talks collapsed! Again, any suggestion of coming to terms with the Mau Mau was bitterly resented by many of the settlers, some of whom went so far as to counter the dropping of official surrender pamphlets by dropping some anti-surrender pamphlets of their own from private planes!

These three attempts at arranging mass surrenders demonstrate some of the difficulties involved. It is a matter of careful judgment as to timing and the terms offered; both sides inevitably doubt the sincerity of the other and suspect treachery. The Public Relations aspect is also difficult, since all the facts and factors involved cannot usually be explained to the general public, who are therefore liable to wax indignant on the basis of false assumptions.

In the Kenya Emergency, the problem was not so much the usual one of whether the guerrillas would use the truce negotiations as a respite to reorganize; it centred more on the long-term implications of coming to terms with the Mau Mau. It was felt, particularly by some settlers, that in dealing with a revolt such as Mau Mau the only possible policy was 'unconditional surrender', in order to ensure that this evil influence was utterly destroyed. This may have been a valid argument with regard to the hard-core insurgents, but the surrender offers were aimed more at the less dedicated rebels and the waverers, who might surrender if they felt that they could find the opportunity and that it would be worth their while to do so. These were the people whom it was hoped to win over in the long term and to convert to better ways. It was argued that this was more likely to be achieved by reasonably generous terms, such as were offered, rather than by insisting on 'unconditional surrender'.

Allegations of Brutalities

The Emergency in Kenya led to a series of allegations of brutality by the Security Forces against Mau Mau prisoners and detainees. Certainly there were some proven cases and it would be naïve to pretend that there were not other cases which never came to light. Violence does breed violence and the Mau Mau atrocities were unbelievably brutal, obscene and pointless; furthermore, the lawful powers available to the police and others for obtaining much needed information must at times have seemed very inadequate.

But there was never any official condoning of brutalities, and any excesses that did occur were firmly condemned and punished by both the civil and military authorities. One of General Erskine's first acts on arrival was to order that any alleged cases of illegal action should be reported to him personally. The Government set up a Ministerial body to investigate all such complaints, and the Commissioner of Police ordered the C.I.D. to investigate any allegation made against the police. Some settlers, however, believed firmly in giving the terrorists a taste of their own medicine and they resented being expected to fight a savage insurgent organization using only Queensberry rules. They even went so far as to set up a fund to pay the legal expenses of those convicted of offences 'in the course of their duty'.

Unquestionably the authorities were right to set their faces resolutely against any lowering of the well-established standards of British justice. But it was not an easy principle to maintain, for the Mau Mau were quick to appreciate how this attitude could be turned to their own advantage. They soon realized that all allegations of brutality would be fully investigated by the authorities and that the person concerned would be suspended from active duty while inquiries were carried out. They could therefore all too easily arrange to rid themselves, temporarily at least, of any Police Officer who was proving particularly effective against them; they merely made indignant allegations that he had beaten or tortured to death a Mau Mau suspect—and they would even produce suitably mutilated bodies to support the charges.

Intelligence

Intelligence in Kenya prior to the Emergency came from two main sources, Special Branch and Government officials, and much useful information was provided from both. But to be of value, intelligence must not only be collected; it must also be collated and the deductions made from it must be disseminated to the right people in the right form, so that appropriate action can then be taken.

Mr. Corfield in his Report is clear that there was failure in this respect both during and before the Kenya Emergency.[1] There were two main weaknesses. Firstly, there was no one body which was in a position to analyse and evaluate the plentiful factual information that was produced on the subject of Mau Mau between 1946 and 1952; as a result, no constructive deductions were made and there were no recommendations as to steps to be taken. Secondly, there was no individual who was either qualified or in a position to ensure that this was done, and to see that intelligence appreciations were disseminated to the right quarters and were acted on. It was in fact one of the many responsibilities of the Attorney General; and, personalities apart, this arrangement did not work out.

These weaknesses were soon realized once the Emergency broke, and urgent action was taken to remedy them. But the usual problem then arose of integrating the civil and military intelligence organizations. There was a requirement for two types of intelligence, the strategic, which covered trends, developments and long-term plans, and also the tactical, which formed the basis of all day-to-day security operations.

The Army and the Police were concerned with both types, but there was no joint intelligence effort to meet their needs. It had to be built up from scratch, and General Erskine made a start by integrating his intelligence staff with that of Headquarters Special Branch. But there was still no single Chief of Intelligence, with the duty of collecting, co-ordinating and disseminating all available information on behalf of the War Council. This lack of effective integration of intelligence was not confined to the higher levels but persisted throughout the chain of command.

It is easy to use hindsight, but it is evident now that the establishment of an effective joint intelligence organization at an early stage might have done much to defeat, and possibly even to prevent, the Mau Mau insurgency.

Large- and Small-scale Operations

Counter-insurgency campaigns call for large-scale operations

[1] *The Corfield Report*, Ch. III.

in the form of 'sweeps' and 'cordons and searches' by perhaps several battalions worth of troops; there is equally a requirement for patrols, ambushes and infiltration tactics by small groups. There is a time and place for each, and the mixture will vary with every campaign; the skill lies in assessing the balance between them, and the Kenya Emergency provides a useful illustration of this problem.

Large-scale operations are often the only offensive action open, if there is insufficient information available to permit of pinpoint attacks against specific targets. There will also be occasions when it is best to tackle a large area all at once, and the 'cordon and search' of Nairobi is a good example of this. Large-scale 'sweeps' do not, however, often kill many guerrillas, though they can be of value by harassing them and by forcing them to move from familiar to strange ground. But they are costly in effort and, unless they are successful, they do not enhance the reputation of the Security Forces with the populace.

Small-scale operations in the form of constant, aggressive patrolling, ambushes and 'pseudo-gangs' pay excellent dividends for the amount of effort expended, but they depend on two essential factors, highly trained troops and good intelligence. Neither of these requirements has usually been available in the early stages of most past counter-insurgency campaigns, and they were certainly not present initially in Kenya.

The large sweeps through the Kenya forests may well have been the best techniques against the Mau Mau in the early phases of the campaign but it is questionable whether they were still justified after Operation 'Anvil' had provided so much intelligence on the Mau Mau Order of Battle. Operation 'Hammer' in January 1955 by nine battalions resulted in the elimination of only 161 Mau Mau guerrillas. Yet, one 'white hunter', who used to track the guerrillas into the forest as if they were wild game, killed over 100 of them single-handed. It is the difference between trying to catch fish in a weedy pond by splashing about with a rather wide-meshed net as opposed to adopting the tactics of the pike, and lurking quietly in the weeds ready to snatch unsuspecting fish as they swim by.

With the advantage of hindsight, it does seem that, although the large-scale operations before Operation 'Anvil' were probably justified, a switch to small-scale tactics at an earlier stage would have paid off, and this was confirmed by the striking success of the 'pseudo-gangs' once they were instituted.

Support of the Populace

The final point of interest is the changing attitude of the local population during the campaign. The Kikuyu, who were the only tribe really concerned in the insurgency, began by being generally loyal to the Government, but were rapidly induced by the intimidation policy of the insurgents to transfer their support, at least passively, to the revolt. But when it became evident that the Security Forces were gaining the upper hand, the waverers began to jump on to the Government bandwagon.

The Government did win the support of the local population in the Kenya Emergency, and without offering any political solution. It was achieved firstly by isolating the basically loyal Kikuyu from the Mau Mau through resettlement and by the formation of the Home Guard. Secondly, these same measures also made it possible to protect the loyal elements and the waverers from the insurgents and so made intimidation measures less effective. Finally, but not least important, obvious successes were gained against the Mau Mau, notably in 'Operation Anvil'; once this was achieved, confidence in the Government grew and the grip of the Mau Mau weakened.

There was relatively little insurgent political activity alongside the military campaign in Kenya, largely because the Mau Mau lacked any political leader to fight their cause for them and to rally public opinion outside Kenya to their cause. For the same reason there was relatively little propaganda effort by either side, and it remained a colonial problem instead of being made into an international issue.

From the Communist viewpoint the Emergency must have been regarded as a satisfactory entry on the credit side of the cold war scoreboard. It had been a distinct embarrassment to Britain militarily, politically and ideologically. No effort or intervention had been required by the Communists themselves beyond moral support and encouragement, and yet the long-term advantages to them were considerable, in that it hastened the end of British rule in Kenya and opened up possibilities for spreading the cold war into Africa, which the Communists have not been slow to exploit.

Map II. KENYA, 1952–56

The Emergency in Cyprus (1955-1959)

Chapter 12

CYPRUS—PHASE ONE (1954–55)

Cyprus was ruled by Turkey for 300 years until 1878, when the Turks agreed that the island should be placed under British administration in return for British protection of Turkey against Russia. When Turkey entered the First World War as an ally of Germany, Britain annexed Cyprus; this action was recognized in 1923 by both Turkey and Greece in the Treaty of Lausanne, which left Britain with full and undisputed sovereignty over the island.

Cyprus has never belonged to Greece, but 80 per cent of the population of 524,000 are Greek-Cypriots,[1] speaking Greek and belonging to the Greek Orthodox Church. A further significant factor is that the secondary education in Cyprus was run on a racial basis, with separate schools for Greeks and Turks, rather than on a Cypriot national basis. As a result, the Greek-Cypriot schools based their teaching on the Greek curriculum—which encouraged sympathy for Enosis, or Union of Cyprus with Greece; indeed some of the maps and atlases showed the island of Cyprus as 'unredeemed Greece'.[2]

Cyprus is only 3,572 square miles (about half the size of Wales), but it has long been considered of strategic importance to Britain and her Allies in the control of the Near and Middle East; it is of significance that in the early 1950's it was considered to be essential to retain 'Cyprus as a base' rather than being satisfied with

[1] Population figures for 1955:

	Greek-Cypriot	419,000
	Turkish-Cypriot	105,000
	Total	524,000

[2] *Cyprus, the Dispute and Settlement*, Chatham House, June 1959, p. 11.

'bases on Cyprus', as was to be accepted eventually. The former view persisted until the Suez crisis, when there was a reappraisal of our essential needs on the island.

Cyprus in 1954 was a British Colony ruled like Kenya by a Governor (then Sir Robert Armitage) on behalf of the Colonial Office, with a Legislative Council consisting of nominated officials; there were no elected representatives of the local population in the Government. The island was divided administratively into Districts, each with its District Officers as in Kenya and Malaya; but with the difference that less than a quarter of them could speak the local language.[1]

The police were organized and trained for normal peacetime duties, but were poorly paid and equipped and were certainly in no state to take on a fierce insurgent organization. Little effort had been made to improve their efficiency against such a contingency, for it was generally assumed that this tiresome problem of Enosis would be settled by negotiations in due course.

Enosis was no new problem to Cyprus; it was first mooted in the middle of the nineteenth century, and its chief exponents have always been the Greek Orthodox Church in Cyprus. This is not as outrageous to the Greek or Cypriot mind as it is to the British, for the power and influence of the Church in Cyprus have always been such that the people have looked to their bishops for temporal as well as spiritual guidance. Archbishop Makarios was therefore only following a well-established precedent, when he became the chief exponent of Enosis; he was the Archbishop, but he also bore the title of Ethnarch, which signifies the acknowledged political leader of the Greek-Cypriots.

The movement for Enosis made little impact until after the First World War; it was then raised periodically, but with little effect, until 1931 when there were violent disturbances, which led to the burning down of Government House in the capital, Nicosia. As a result, the Government took stern retaliatory measures, including the dissolution of the Constitution and of all elected bodies; after this, the question of Enosis was not raised again for the next twenty years. But it had not died.

[1] *Cyprus, the Dispute and Settlement*, Chatham House, June 1959, p. 15.

On 15th January 1950, a plebiscite on Enosis was organized by the Church; signatures were collected actually within the churches and it was claimed that 95·7 per cent of those entitled to vote had supported Enosis.[1] On 20th October of the same year, the brilliant thirty-seven-year-old Bishop Makarios of Citium was elected Archbishop Makarios III of Cyprus, and promptly declared: 'I shall not rest for a moment in my efforts to see Union with Greece achieved.'[2]

He soon matched his words with deeds when in 1951 he initiated two significant moves. First, he invited Colonel Grivas to return to his birthplace, which he did in the same summer; ostensibly, his visit was to build up PEON, the Pancyprian National Youth Organization, but it was in fact a reconnaissance for the impending 'struggle for independence'.

Secondly, Archbishop Makarios set out to induce the Greek Government to give public support to Enosis. This was not so easy, and it called for all his skill in diplomacy and intrigue; Greece had no wish to quarrel with Britain, who was an old friend as well as a wartime ally and a partner in NATO. Yet in May 1954, the Greek Prime Minister, Field Marshal Papagos, announced his intention of raising the issue of Enosis at the United Nations, and demanding its implementation.

Two events served to strengthen him in this decision, apart from the pressure exerted by the Ethnarch. In June 1954, Britain had announced that Headquarters Middle East Command would move from the Canal Zone to Cyprus. Then in August, a statement was made by the Attorney General in Cyprus to the effect that existing anti-sedition laws would now be enforced. This was interpreted in Cyprus as forbidding all reference to Enosis, and, on this basis, aroused widespread resentment. The local newspapers held a one-week strike and Archbishop Makarios defied the law by preaching on the subject in church.

In December 1954, the United Nations duly considered the Cyprus question, but decided to defer any action on the Greek resolution indefinitely; this ruling was the signal for riots in both

[1] *Cyprus, the Dispute and Settlement*, Chatham House, June 1959, p. 11.

[2] *Cyprus and Makarios*, Stanley Mayes, p. 21.

Greece and Cyprus, and at the same time, Athens Radio opened up a virulent propaganda campaign against Britain and Turkey, which was to continue until the end of the Emergency. Meanwhile children at Greek-Cypriot secondary schools were being taught that 'a new War of Independence' was beginning, which demanded their full support.

The first undeniable act of the insurgency came not long after, when, on 25th January 1955, the authorities seized a Greek ship called the *Ayios Georghios* (or the Saint George!) when it landed near Paphos.[1] The captain and Greek crew of five were arrested, as well as a reception committee of eight Greek-Cypriots, who were awaiting the vessel by night at a secluded cove. On board were found arms and 10,000 sticks of dynamite.

'Yes,' the ship's Captain was alleged to have declared. 'We brought dynamite to free Cyprus.'

The preliminary hearing of those arrested was held on 16th March, and reference was made for the first time in public to a subversive organization established to fight for Enosis. It was at this time called EMAK (National Front for the Liberation of Cyprus), but it was not heard of again under this title, and in its place sprang up the National Organization of Cypriot Combatants, better known as EOKA.

EOKA burst on Cyprus and on the world on 1st April 1955, when the whole island rocked to a barrage of explosions, which damaged the Secretariat, the broadcasting station, several police stations and other official buildings; there were no casualties (except one saboteur who blew himself up) but some £60,000 worth of damage was caused, and attention was drawn to Cyprus worldwide. Amid the wreckage were scattered leaflets signed 'Dighenis', a romantic Robin Hood figure of ancient Greece, whose name was now to be taken by Colonel Grivas as the acknowledged leader of EOKA and of the 'struggle for liberty'. He had landed in Cyprus secretly on 10th November 1954, and had been in hiding ever since.

These events brought Cyprus into the world headlines, and also made it clear that Britain could no longer continue to ignore the

[1] See map, p. 154.

issue of Enosis. The Colonial Secretary, Mr. Lennox Boyd, visited the island in July and met Archbishop Makarios for the first time; he then convened a conference in London in August to be attended by representatives of both Greece and Turkey. But both these countries had by now adopted irreconcilable standpoints from which they were not prepared to budge.

Greece demanded unconditional Enosis, on the grounds that 80 per cent of the population favoured it. Turkey opposed Enosis equally firmly on the grounds that it was unfair to the Turkish-Cypriot minority; and in any case, they argued, if Cyprus was going to be given away to anyone, it should be given to Turkey as a former owner rather than to Greece, who had never occupied the island.

Britain for her part was in possession of the island by right of treaty, and intended to stay there. Moreover, it was argued that the use of the island as a strategic base was essential not only to Britain but also to NATO. At the same time, it was realized that it was important to avoid a quarrel between the three members of NATO who were involved, for dissension might have disastrous repercussions throughout the Balkans, Europe and the Middle East; such a development would benefit only the Communists, who were, as usual, waiting to exploit the situation in any way they could.

The Greek-Cypriots themselves were unhappy, for although they believed devoutly in Enosis, they had no real quarrel against the British people, and indeed, they fully appreciated the benefits of British rule and trade in Cyprus. There was much genuine goodwill and a desire for a peaceful solution, for the Greek-Cypriots are by nature a passive and peace-loving nation (as Grivas himself complained) with little desire for violence if other ways to Enosis could be found.

As was to be expected, the opposing viewpoints proved irreconcilable, and the talks duly broke down on 7th September. Two weeks later, the United Nations announced that it would not now put the Cyprus question on its agenda at all; tension immediately mounted on all sides, and it was all too clear that Britain must be prepared for trouble in Cyprus.

EOKA had meanwhile not been passive, and there were constant acts of sabotage and attacks, chiefly directed against the police force. Efforts were also made to subvert or intimidate not only the police, but also Cypriot officials in the Government and to divert their loyalty towards Enosis. The public, who by nature believed in keeping out of trouble as far as possible, now became increasingly reluctant to give information to the authorities, even when a policeman was murdered in daylight in a crowded street.

Security precautions were tightened up, and in June some Emergency Regulations were passed, chiefly to give the Government the power to detain suspects without trial. A new phase began when the appointment was announced on 25th September of a new Governor of Cyprus, who was to be Field Marshal Sir John Harding, K.C.B., D.S.O., M.C., a former Chief of the Imperial General Staff.

The Field Marshal arrived on 3rd October, and immediately set to work on three tasks. Firstly, he reorganized the Administration, the Services, and the Police into one effective, co-ordinated Security Force, geared to the restoration of law and order. At the same time, he opened negotiations on a man-to-man basis with Archbishop Makarios. Finally, he intensified operations against EOKA.

As General Templer had done in Malaya, the new Governor himself assumed the appointment of Director of Operations as well as Governor, and thus had the most effective possible control over all aspects of Government. To assist him, he made the Deputy Governor responsible for the administrative aspects of government, while Brigadier Baker was appointed as his Chief of Staff, to be directly responsible to him for executive control of all Internal Security operations.[1]

Under the Chief of Staff was set up a joint staff of all three Services, together with a joint intelligence organization; another important step was the appointment of an Under Secretary (Internal Security), who was to be directly responsible to the Gover-

[1] The title was changed on 1st January 1957 to Director of Operations, and a Chief of Staff was then appointed to be responsible to him.

nor for all civil aspects of security operations and for liaison between civil and military. One Joint Operations Centre was established and manned twenty-four hours a day, thus ensuring effective co-ordinated control over day-to-day operations.

A simple command and control structure (see *Appendix E*), was set up, giving the Governor, through his Chief of Staff, a close control over all aspects of the Emergency. He had direct contact with his military commanders as well as with the Police, Intelligence Information Services and all civil Departments, and was himself represented in discussions with all these bodies by his Under Secretary (Internal Security).

Because Cyprus was so small in area, there was no elaborate hierarchy of States, Provinces or Regions, and the Governor was able to deal directly with the District Security Committees without any intermediate headquarters. These Committees would meet normally about every other day, but more frequently if required, and would on occasions have meetings with the Governor himself.

This organization was successful in its primary aim of establishing an integrated 'civil, military, police' machine for the control of the Emergency. The significant feature is that the machinery was set up *before* the Emergency was declared, instead of being improvised afterwards, and it in fact lasted with virtually no change to the end.

Other elements of the Government also needed and received attention. The police were barely coping with the increasing lawlessness, and it was clear that with their very inadequate resources of manpower and equipment they must be given support urgently. As a first step, 300 additional police officers were drafted in, raising the total to 2,000 (which can be compared with a strength of 4,911 two years later). Colonel Geoffrey White, Chief Constable of Warwickshire, arrived as Commissioner of Police, and his energy and fine leadership did much to build up a new, revitalized Force and to weld it into an efficient striking force alongside the Army.

This, however, took time, and in the meantime, the Army helped by taking over some police duties such as patrols and guards on Vulnerable Points, as well as reinforcing many police stations, which had become a popular target for the insurgent attacks.

Information was sadly lacking at this stage, and there was no real co-ordination of police and military intelligence staffs; nor was existing intelligence being fully collated and disseminated. The Governor therefore appointed a Chief of Intelligence to be responsible to him for the co-ordination of all intelligence sources and resources, a move which paid good dividends before too long.[1]

The Information Services received much the same sort of treatment. Eighteen months earlier, the writer Lawrence Durrell had been appointed Public Information Officer for Cyprus, and he described his first visit to his office as 'giving the impression of entering an abandoned barber's shop'.[2] The new Governor soon decided that much greater emphasis must be placed on Public Relations, and that the Information Services must be expanded and improved. He set up a P.R. Department under a Director General of Information Services, and also appointed a Public Relations Officer for the Police. A Central News Room was opened up, which was soon to be manned seventeen hours a day like an Operations Centre.

Even while he was preparing like any good commander for the 'worst case', the Governor was also seeking a political solution to the showdown which he could see looming ahead. He approached Archbishop Makarios as the accepted leader of the Greek-Cypriots, and tried to reach some understanding with him. But no acceptable basis for negotiations could be found, and after eight meetings, it was evident that the talks had finally reached a deadlock. In October, the Governor flew to London for further talks with Ministers.

EOKA had meanwhile been continuing their insurgent activities, and by mid-November, terrorist attacks by EOKA had resulted in some 30 killed and 291 wounded, as well as damage to 1,260 buildings. Such a situation could not be allowed to continue,

[1] *The Cyprus Emergency* by Brigadier Baker. Journal of the Royal Artillery, April 1958.

[2] *Bitter Lemons* by Lawrence Durrell, p. 152.

but increased efforts by all available Security Forces showed no signs of reducing the threat. All attempts at a peaceful solution had also failed and strong counter-action against the insurgents seemed the only answer. Every possible preparatory move that could be achieved in the time available had been taken, and the seconds were out of the ring.

On 26th November 1955, the Governor declared a State of Emergency in Cyprus.

Chapter 13

CYPRUS—PHASE TWO (1955–57)

The Cyprus Emergency differs from the pattern set out in Chapter 3 and also from the Emergencies in Malaya and Kenya, in that the phase of preparation, in which the Government normally passes from a peacetime to an operational footing, had been largely completed before the State of Emergency was declared. 26th November 1955 therefore saw a deliberate advance to battle by two protagonists, who had already confronted each other across a conference table, and who were now both prepared for a tough, and if necessary, long struggle.

It was indeed to be a Homeric conflict, fought simultaneously in the political and military spheres to an extent which was not the case in either Malaya or Kenya. The Governor represented Britain in both fields, facing Archbishop Makarios in the political arena, and Colonel Grivas and EOKA on the insurgency front. It is an interesting study also to analyse the relationship between Archbishop Makarios and Grivas and to see how the policy and strategy for the campaign were co-ordinated between them.[1] A study of the active and quiet periods of EOKA operations, including three 'truce offers', shows how the political and military aims were interrelated.

Colonel Grivas was a clever and brave soldier, who had made a thorough study of the theory of guerrilla warfare as well as having gained personal experience during the civil war in Greece in 1944–45.

[1] 'I will not lay down my arms unless you yourself ask me to do so. . . .' Letter from Grivas to Makarios, 29th January 1955, and others, as set out in *Terrorism in Cyprus*.

He was inspired by personal ambition, but also by a fanatical belief in the cause for which he fought, which was the 'freedom' of his native land and its union with Greece. He was an impressive leader, who earned the respect and fired the imagination not only of EOKA but also of many of the Greek-Cypriots. He was determined and ruthless, and did not hesitate to condemn to death both insurgents and his fellow Cypriots, if he considered it was necessary for the good of the cause.

His plan was ambitious in that he aimed not to seize and occupy any territory, but to rely on guerrilla warfare and terrorism over a long period to make the British position in Cyprus untenable. In this way, he hoped to force Britain to accept the political solution of Enosis as the only possible alternative to a stalemate and indefinite insurgency. In other words, he hoped to achieve the classic guerrilla objective of undermining his enemy's 'will to fight on' to the point where Britain would decide that their counter-insurgency effort was no longer worth while.

This aim is normally unlikely to be achieved against a firmly established Government purely as a result of insurgency; it requires to have political support, which can use the insurgent struggle to gain a favourable political solution. Grivas had this support to the greatest possible extent. He was backed not only by a brilliant and determined leader in Cyprus, but also by the full diplomatic support of Greece. He could in addition expect substantial sympathy for his cause from world opinion, which was a powerful weapon in his hand, and one which the Communists in Malaya and the Mau Mau in Kenya had lacked.

Despite this support and the sympathy of the local population, Grivas did not, however, achieve his ultimate aim of Enosis. His failure was not the result of military defeat, but due to the decision of the political leadership to accept a political solution based on the Greek and Turkish Premiers' plan for an independent Cyprus. The political aim for which he was fighting disappeared in 1959, and he then had no option but to accept the order from Archbishop Makarios to cease the military struggle.

The methods that Grivas chose to achieve his aim were the obvious ones in the circumstances. He had the Troodos mountains and forests as a guerrilla base, and he also had a sympathetic populace; his guerrillas were therefore able to 'live like fish in water' among the locals in both town and country, knowing that they were not likely to be betrayed so long as the people supported their cause.

EOKA was organized along normal guerrilla lines.[1] The Headquarters consisted of Colonel Grivas and his immediate followers, who numbered less than a score. They lived initially in the Troodos Mountains, hidden in the network of caves existing there, though later, when the pressure against this mountain base increased, they moved into hiding elsewhere. From his headquarters, wherever it was, Grivas organized and led the campaign through an elaborate system of agents and couriers, who kept him in touch with his Regional and local commanders. The supporting elements were recruited largely from the villages, where they were in a position to supply food, shelter, money and information to the operational groups.

The 'active wing' of EOKA consisted of some seven 'mountain groups' each of five to fifteen men, while in the towns there were some fifty 'town groups', which totalled, according to Grivas,-only 220 men altogether. This added up to a total 'hard core' of less than 300, but it was supported by a 'passive wing' of sym[2] pathizers and also by some 'village groups', who were a paramilitary force armed with shotguns.

The 'mountain groups' acted as guerrillas, hiding in the caves on the forested Troodos slopes, and emerging to carry out swift, vicious raids and ambushes in the surrounding countryside.

The 'town groups' had the task of sabotage and terrorism for which they were divided into 'professionals' who formed squads for shooting, bomb-throwing or sabotage, and the 'amateurs' who were used to distribute leaflets, collect information and act as couriers. An example of this division of duties is that the leader of a 'killer group' did not carry his own weapon when on a mission, but would be given it at the vital moment by a 'carrier',

[1] Appendix C, column (*h*).

[2] *Grivas Memoirs*, p. 66.

usually a young boy or girl, who would then take it back after the fatal shot had been fired.

All supporters of EOKA were bound by an oath, based, like the milder Mau Mau oaths, on religion but without any of the vile, primitive ritual associated with the latter. Nevertheless, it was an uncompromising commitment:

> 'I swear in the name of the Holy Trinity that I shall work with all my power for the liberation of Cyprus from the British yoke, sacrificing for this even my own life. . . . I shall never reveal to anyone any secrets concerning our Organization, neither the name of the Leader nor that of any of the members, even if I am caught and tortured. . . . If I betray this oath, I shall deserve every punishment meted out to me as a traitor and may I suffer eternal shame.'

Such an oath, administered in the name of the Holy Trinity, was not to be lightly disregarded; and to make quite sure of its effectiveness (in case divine retribution did not prove quite adequate!) the punishment for breaking the oath was indeed death, and it was meted out ruthlessly on the direct orders of Grivas.[1]

Grivas began by waging a campaign of both terrorism and traditional guerrilla warfare in the form of 'hit and run' attacks directed primarily against the police. Using the Troodos area as their base, his mountain gangs descended on to the plains and villages to lay ambushes, attack Police Posts and to sabotage installations. This policy, however, proved vulnerable to counter-measures taken by the Security Forces, who aimed at isolating and then attacking the mountain bases, which were too small in area to give the guerrillas proper mobility and security.

Grivas therefore changed the emphasis (after Operation 'Pepperpot' in 1956)[2] to terrorism in the towns, carried out by the gangs based there. Guerrilla warfare continued, but in a secondary role only, for the Security Forces never again allowed EOKA sufficient mobility to resume their 'hit and run' tactics.

[1] 'I informed Thalis to murder Kyriakos on . . .'. Letter by Grivas, 16th June 1955, *Terrorism in Cyprus*.

[2] P. 133.

Grivas intended that his strategy should achieve three results. Firstly, it would keep Cyprus and the cause of Enosis in the world headlines, and if cleverly exploited by propaganda, it would also present Britain in an unfavourable light. Secondly, he hoped to drive the Government to adopt repressive measures against the insurgents, which would also cause hardship to the populace, and so increase support for EOKA and Enosis. Finally, he was determined to ensure that EOKA gained the support of the local population, if necessary by intimidation. In the course of the campaign, he achieved considerable success with this plan, although he did not achieve his ultimate objective of Enosis.

EOKA celebrated the declaration of the Emergency on the very same day that it was announced. At a Caledonian Ball held in the Ledra Palace Hotel in Nicosia, a grenade was thrown within a few feet of the chair on which the Governor would have been sitting, had he not cancelled his attendance at the last moment as being inappropriate in the circumstances.

A further attempt was made on the Field Marshal's life in March 1956, when a Cypriot servant employed at Government House managed to smuggle in a time bomb concealed in a belt round his waist. He was then able to place it undetected under the mattress on the Governor's bed. Fortunately, he had not allowed for the fact that the Field Marshal always slept with his window open, and the cold night air slowed up the time pencil, so much so that the bomb did not explode until after it had been discovered the next morning. The Governor declared that he had slept extremely well!

On 13th December 1955 the Communist Party in Cyprus, known as AKEL, was proscribed and 120 suspected members were detained. The reason given was that they were opposing and hindering negotiations for a settlement in Cyprus; this was an activity in keeping with their policy of stirring up dissension wherever an opportunity offered, and Cyprus was a promising target. They had, in April, condemned violence as a means of settling the problem of Enosis, but this did not mean that they were opposed to Enosis or that they supported Britain; it was merely a shrewd tactical move, which left them free to make what capital they could out of the situation. In the course of the Emergency, the Communist bloc did its best to cause dissension between the nations involved in the Cyprus dispute, and so weaken the NATO alliance.

For the first few months of the Emergency, the Security Forces were on the defensive, and priority had to be given to tasks such as supporting the police and guarding Vulnerable Points in order to maintain essential services, rather than hunting insurgents. Efforts were made to protect the populace against intimidation, but with limited success only.

The Security Forces had already scored one good success, however, in their first direct encounter with EOKA. On 15th December 1955, four leading members of the insurgents were sitting in ambush on a road near the village of Mersinaki, waiting for a military vehicle to attack. They chose a Land Rover and opened fire, killing the driver; but the officer sitting beside him coolly took the wheel, steered the vehicle into the ditch and began to return the fire. When his ammunition was exhausted, he crawled back to the vehicle under heavy fire, seized his driver's Sten gun and continued the fight. The guerrillas tried to trick him by pretending to surrender, but he spotted the hidden machine-gun covering him and fired a burst at it, which killed one man and wounded another, who fled; the remaining two then surrendered.

The dead guerrilla was Charalambous Mouskos, a cousin of the Archbishop, and a man with a price of £5,000 on his head. His funeral was turned by the Cypriots into that of a national hero and martyr; Archbishop Makarios himself conducted the service, after which large crowds filed past the bier.

The British officer was Major Brian Coombe of the Royal Engineers, and he was awarded the George Medal for his bravery. He was persuaded to give a Press conference, but brought proceedings there to a premature close by a short speech, in which he said: 'I think you gentlemen are barking up the wrong tree. . . . When my driver was killed beside me, I felt it was my duty to bring his murderers to justice; that is why I killed a frightened young Cypriot. It seems to me tragic that the misguided people of Cyprus should honour this man as a national hero . . . and that the British were exultant that I had killed a Cypriot. I want to make an appeal to you. Do not deepen the rift between the Cypriots and the British. Do not encourage the Cypriots to build up a hero by producing a British hero; do not force them to make a hero of a murderer. I ask that there should be no jubilation, no exulting over this affair.'

There was none, but there was to be yet one more twist to this symbolic incident. The two guerrillas who had surrendered, Zakos and Michael, were duly tried and condemned to death. But on 4th August 1956, a few days before their execution was due, EOKA kidnapped a seventy-eight-year-old Englishman, Mr. John Cremer, who had lived on the island for many years. They claimed that he was an intelligence agent, and threatened to kill him if the two condemned men were hanged. But Zakos then sent a message to Grivas saying that he accepted his death sentence as just, and appealing to Grivas to spare Cremer; Grivas agreed and Cremer was set free unconditionally. The Governor then carefully reviewed the death sentence on the two guerrillas, but decided that the Cremer incident was not relevant to the administration of justice in the case of the two condemned terrorists, and they were hanged.

Their fate provided a good example of the enigma of the Greek-Cypriots and their attitude to EOKA. The dead terrorists were loudly proclaimed by the Greek and Cyprus propaganda machines as heroes and martyrs to the cause of Enosis, and it was asserted that the whole population regarded them as such. The Cyprus Government maintained, however, that any demonstrations or expressions of sympathy for the men were far from being either sincere or spontaneous, but were only the result of fear, because no one dared to differ from the official EOKA line. These differing interpretations were to recur regularly throughout the Emergency, and were the cause of much controversy as to the best policy to be adopted in handling the local populace.

EOKA did not normally take hostages, but in December 1955, two soldiers were captured and executed, their bodies not being found until about a year later. This incident aroused considerable indignation and horror, and it seemed possible that it might even turn some of the Greek-Cypriots against EOKA; but there was no apparent change in their attitude. Certainly no further hostages were executed, and it is said[1] that Grivas ordered that there should not be, for fear that he might otherwise alienate the populace, which was the one thing he knew he had to avoid at all costs.

In February 1956, 16th Independent Parachute Brigade was moved to Cyprus, and this enabled the Governor to consider more offensive action against the insurgents. He was still obliged to employ troops on defensive tasks such as reinforcing the police, protecting communications and guarding installations vital to the island's economy. But it was appreciated that it might be possible to take action to isolate the 'mountain groups', since the area of their base in the Troodos Mountains was limited and could perhaps be cut off from the surrounding country.

In June 1956, a major 'cordon and search' operation called 'Pepperpot' was mounted against the Troodos area. It involved a vast encircling movement from three sides, followed by a series of intensive searches and interrogations, aimed at the capture of Grivas himself, who was known to be within the cordon. Although the guerrillas had known that the operation was imminent, they were deceived as to its direction and form, and were undoubtedly taken by surprise.[2] Kykko Monastery was occupied and evidence of its use by the guerrillas was found. During the first few days, twenty known guerrillas were captured and the net closed tighter on Grivas; three times he was all but caught, but finally he managed to evade the trap, and became thereby an almost legendary figure to the Cypriots, who now called him 'Apiostos', or 'the one who cannot be taken'.[3] He had, however, lost two complete 'mountain groups' and he decided that the mountains were no longer safe for him personally as a hiding-place. The operation was a definite success for the Security Forces, in that it seriously disrupted the activities of the guerrillas in the mountains, and forced Grivas thereafter to resort to terrorism, with guerrilla warfare in a secondary role.

It was during this operation that a fierce, uncontrollable forest

[1] *Cyprus Guerrilla* by D. Alostos.
[2] Ibid. [3] Ibid.

fire broke out in the search area; twenty-one British soldiers were burned to death, many vehicles were destroyed and the search was disrupted. EOKA promptly claimed that the fire had been started deliberately in an attempt to smoke out Grivas, and that it served the British right that they themselves were caught by it.

The next move in the struggle came from the British Government, who sent Lord Radcliffe out to Cyprus in May 1956 to work out a new consitition for the island. While he was still at work EOKA made their first truce offer; on 16th August, pamphlets appeared signed 'EOKA—The Leader', and promising an immediate cease-fire 'to test Britain's sincerity'. A strange silence descended on the island, and the populace began to celebrate what they hoped might be a lasting peace.

The truce was taken by the Government, however, as evidence that EOKA was sufficiently hard pressed to want a respite, and that full use should be made of this development to test their sincerity too. It was therefore decided that there should be no negotiations with the insurgents, but they were offered three weeks in which to surrender with their arms; they would then either be imprisoned or deported to Greece as opposed to being hanged, which would be their fate if captured otherwise.

EOKA's response was defiant, and seemed to indicate that they were not near defeat and regarded the terms offered with scorn. 'Victors do not surrender', declared the pamphlets, 'Come and get us', and the explosions began again. The population as a whole also seemed to reject the offer, and did not show any signs of being won over to support of the Government. It was considered by some observers that more generous and diplomatic terms might have succeeded where the ultimatum which almost amounted to 'unconditional surrender' failed. Even with hindsight, it is still hard to tell.

Their offer rejected, EOKA renewed their campaign of violence, and in one week in September, thirteen people were killed and twenty-five injured. A bomb attack in Nicosia killed two Servicemen and two Policemen, and the Old Town was put under a nine-day curfew. On 6th September violent anti-Greek riots broke out in Istanbul, and this led to an increase in tension between Greek and Turk in Cyprus. Hatred and violence were at a peak, when a new situation intervened—the Suez crisis.

16th Independent Parachute Brigade was withdrawn from Cyprus and the Army found itself with other problems to think of in addition to EOKA. Grivas seized his opportunity, and during November made 416 attacks, the highest monthly figure yet; forty people were killed, several RAF planes were blown up and widespread sabotage was carried out, but he caused no interference with the mounting of the Suez operation.

December 1956 saw the publication, and also the rejection, of the Radcliffe Report, which put forward a possible constitution for the island; a political solution seemed to be as far off as ever and the insurgency continued. The year ended with a total casualty list, since the start of the Emergency, of 196 members of the Security Forces killed and 382 wounded against a known thirteen EOKA killed and an unknown number wounded; 212 guerrillas had been tried before the Supreme Court and eight of them had been hanged.

The year 1957 began well for the Security Forces. With the return of troops from Suez, it became possible to form an independent striking force of brigade strength; a further significant development was the formation of a helicopter force which gave the troops invaluable mobility against the guerrillas in the areas outside the towns. The contents of the Grivas diaries,[1] together with the results of interrogations, the use of informers and patient intelligence had disclosed much of the organization, methods and mentality of the insurgents, and it thus became possible to mount small-scale operations directed against specific targets—with encouraging results. In the first quarter of 1957, EOKA lost sixteen killed, including three prominent leaders, and sixty captured.

On 22nd February, the U.N. called for a resumption of negotiations on Cyprus, and on 14th March, Grivas again offered to suspend operations 'in compliance with the spirit of the U.N. resolution and to facilitate the opening of talks for a settlement, provided that Makarios was released to speak for Cyprus'.

[1] See *Terrorism in Cyprus, the Captured Documents*, published by the Colonial Office in August 1956.

Terrorist activities did cease at once, and Archbishop Makarios was offered his release in return for a public condemnation of violence. Although his 'denunciation' was decidedly equivocal, he was duly freed, on condition that he did not return to Cyprus; he flew to Athens, where he received an enthusiastic welcome and promptly took up the struggle for Enosis once more.

Grivas was offered a safe conduct to Greece, but declined, and the search for him was renewed. The truce was still in existence, however, and there was some relaxation of the security regulations; some detainees were released, and daily life on the island became almost normal, with troops walking about unarmed and fraternizing happily with the locals.

Operations against EOKA continued, however, for the insurgents still formed a threat and also a bargaining counter for Archbishop Makarios in any negotiations. This was shown by a further outbreak of violence in October and November, when Cyprus was again being debated by the U.N.

In November 1957, it was announced that Field Marshal Harding was retiring as Governor and that he would be replaced by the late Governor of Trinidad, Sir Hugh Foot, a man noted for his firm liberal convictions. He had been Colonial Secretary in Cyprus from 1943 to 1945, and so was well qualified for the role of negotiator.

This change meant, for the Security Forces, the end of Phase Two, for their campaign against the insurgents was never quite the same again. The new Governor was to initiate a phase of conciliation and diplomacy, with military action being used only as a last resort if other methods failed. Security operations continued, and the struggle was at times as bitter as ever, but it was to take a secondary place to political negotiations.

Chapter 14

CYPRUS—PHASE THREE (1957–59)

'I want a credit of time,' declared Sir Hugh Foot as his opening gambit soon after his arrival in December 1957, whereupon EOKA promptly distributed leaflets agreeing to give him the chance that he asked for.

He set out immediately to gain the confidence of the local population. 'I will deal with you honestly and sincerely, . . .' he declared, and told them that he was their . . . 'friend and servant', whose desire was to build 'bridges of trust'. Such words may have appealed to the Greek-Cypriots, but they earned the distrust of the Turkish community, who suspected appeasement—at their expense. They began to demand a partition of Cyprus and to talk of a Turkish counterpart to EOKA. Urgent talks took place in London, Athens and Ankara during January and February 1958 but they only emphasized the deadlock; and the danger of inter-communal strife in Cyprus was added to the continuing threat of renewed insurgency.

The year 1958 was a year of incessant political search for a solution that might prove acceptable to all concerned, which now meant Britain, Greece, Turkey, Greek-Cypriots and Turkish-Cypriots. Four times in the year,[1] Harold MacMillan as Prime Minister put forward proposals, only to see them rejected either by Greece or by Turkey or by both. On each occasion, EOKA suspended operations during the negotiations, renewing their attacks when talks failed.

In March, the truce, begun in February, ended, each side blam-

[1] 12th February; 12th June; 16th August; 7th October.

ing the other for provocations. This time EOKA added a new technique to their normal guerrilla and terrorist tactics, and called for 'passive resistance' against the Cyprus Government. A boycott was imposed on the sale of British goods, and was enthusiastically enforced by EOKA supporters; Greek-Cypriot pupils were withdrawn from British schools, and as a final defiant gesture, British street names were obliterated. Attacks on 'traitors' were stepped up, and a new wave of sabotage caused considerable damage.

Tension grew, and in June the patience of the Turkish-Cypriots finally snapped, when it seemed likely that their demands for partition would not be met. Violent intercommunal riots broke out all over Cyprus, and the long-suffering British troops found themselves given the new task of keeping Greeks and Turks apart; and as usual, being accused by both sides of bias towards the other. In June and July, fifty-six Greek-Cypriots and fifty-three Turkish were killed in these disturbances.

On 1st October, the MacMillan Plan of August, which had been accepted by Turkey but rejected outright by Greece and by Archbishop Makarios, was due to be put into effect with Turkish support. It envisaged Cyprus being ruled by a British Governor, flanked by one representative from Greece, one from Turkey, and by six Cypriot Ministers—four Greek and two Turkish; in each town there would be separate municipalities for the two communities.

EOKA decided to demonstrate violently against the Plan and on 1st October there was a general strike. Next day, a British civilian was shot by terrorists, the first such incident since January 1957. On 3rd, Mrs. Cutliffe, the wife of a Royal Artillery sergeant, was shot in the back five times while shopping and was killed; her daughter with her escaped unhurt, but another wife with her was killed. Feelings ran fever high; EOKA denied responsibility for the crime, but few believed it. The Greeks accused the Turks, and the Turks blamed the Greeks. The Governor held Archbishop Makarios morally responsible and EOKA blamed Britain.

During the month of October, EOKA made fifty attacks, killing twenty-three and wounding 300. The Security Forces, now commanded by General Darling, who had arrived on 10th October, took immediate and strong counter-measures, and the grim deadlock continued. A bomb exploded in a NAAFI sofa, killing two airmen, whereupon the Government implemented a measure which they had refrained from carrying out earlier, and dismissed 5,000 Greek-Cypriot employees in NAAFI and other Service establishments. A call was made for 500 volunteers from the U.K. to work on the staff of NAAFI in Cyprus, and there were 18,000 applicants, even though the terrorism continued and Service wives now shopped only at specified hours, guarded by armed troops.

At the end of November 1958 the Cyprus issue came up for its annual review at the U.N., and the customary, not-very-hopeful appeal was made for a peaceful solution. There seemed little to suggest any significant improvement in the situation, for Turkey was still opposed to Enosis as proposed by Greece, and Greece was still opposed to the Turkish suggestion of partition. But in fact, all parties were growing increasingly weary of the continuing violence and bitterness, and were less adamant in their refusal to compromise. Just before Christmas, the Governor reprieved four EOKA killers from the gallows, and then on Christmas Eve, EOKA unexpectedly announced a complete cessation of their activities 'on the understanding that the other side does the same'. The Security Forces did not accept the truce offer and the search for the insurgents continued, but the terrorism did cease.

On 10th February 1959, the Greek and Turkish Premiers, who had been discussing the Cyprus problem together at Zürich, suddenly announced their full agreement on a plan for an independent Cyprus. Britain had not been included in the discussions, but she promptly endorsed the proposal. Nine days later, Archbishop Makarios having joined the conference as the accepted representative of Cyprus, reluctantly agreed at Lancaster House to accept independence for his country, instead of Enosis. Grivas, still in hiding in Cyprus, accepted his master's decision, but even more reluctantly.

On 1st March, Archbishop Makarios returned in triumph to Nicosia, and on 17th Grivas emerged from hiding, had a brief meeting with the Archbishop and was then flown back to Greece.

On landing at Athens, he received a hero's welcome, was promoted Lieutenant-General and granted pay of the rank for life. The EOKA leader was far from happy over the final outcome, for he had failed to achieve his original aim of Enosis but he had little choice and duly endorsed his master's decision that independence must be accepted as the only possible solution. Britain's will to fight on had proved stronger than expected, and it had been Archbishop Makarios not Mr. MacMillan, who had given way first.

The military campaign against EOKA was over, after three and a quarter grim years, and it could only be called a 'draw'. The Security Forces had been remarkably successful in their counter-insurgency campaign in the face of a consistently uncooperative populace. Despite this considerable disadvantage, they had exerted increasing pressure against the insurgents; in particular, the skilful use of helicopters had proved a decisive weapon, and had given the troops such mobility that they were able to neutralize the mountain guerrillas almost completely. They had failed, however, to put a stop to terrorism in the towns, nor had they succeeded in capturing Grivas; and while he was still at large, in command and able to attack, it could not be said that the insurgency was quelled.

The total casualties inflicted by EOKA (as opposed to those incurred in the intercommunal riots) were:

			Killed		*Wounded*
Services			104		603
Police			51		185
Civilians:	U.K.	26		49	
	Greek-Cypriot	203		154	
	Turkish-Cypriot	7		64	
	Others	2		21	
	TOTAL—Civilians		238		288
TOTAL			393		1,076

The casualties inflicted on EOKA are not known for certain;

the recorded total of insurgents killed is ninety, but it was almost certainly higher.

On 13th November 1959, Archbishop Makarios was elected President of an independent Cyprus, with 67 per cent of the votes, and negotiations were then begun over bases on the island. Britain was ultimately allowed to retain sovereignty over base areas amounting to 120 square miles out of the island's 3,572—a far step from the original aim of 'Cyprus as a base'.

On 4th December 1959, the State of Emergency was declared at an end.

Chapter 15

CYPRUS—GENERAL POINTS

Attitude of the Local Population

The attitude of the local population during the Cyprus Emergency was a major factor throughout, and was the key to the policies adopted by both sides. It was evident from the start, to Britain as well as to Archbishop Makarios and Grivas, that success in both the short and the long term depended on the support of the Greek-Cypriots, and each side therefore devoted every effort to trying to win this support for their own cause.

The situation facing Britain was very different to that in the Malaya and Kenya Emergencies; there, they had been faced by a local revolt initiated by only a part of the population and fought out in largely uninhabited jungle, so that the bulk of the community were not directly involved. In Cyprus, Britain faced a national demand for Enosis not only by 80 per cent of the Cypriots, but also by Greece. The campaign was fought by a sophisticated European insurgent force under comparatively urban conditions, so that not only was every citizen involved, but the situation rapidly became an international issue. Finally, the insurgents had a clever and determined political leader also fighting for their cause.

Throughout the Emergency the Greek-Cypriots did give active support to EOKA. The vital question was whether this support was given freely or only out of fear as a result of a policy of intimidation by EOKA. This is a question to which there is probably no conclusive answer even today, and it would require another book to analyse the problem fully. But it is possible, with the advantage of hindsight, to see the problem in some perspective and to draw some conclusions.

There is little doubt that the Greek-Cypriots believed that Enosis was right and proper, and that Britain should agree to it. They were Greeks by upbringing, by religion and by inclination, and their loyalty to Greece was a matter of pride and emotion to them. As a Nicosia truck-driver declared in 1954: 'I would vote for Enosis out of respect for myself, because I am Greek and I am proud of it.'[1]

There was a distinction, however, between voting for Enosis and fighting for it; moreover they fully appreciated the advantages of life in Cyprus under British rule, and hoped that it could perhaps be reconciled somehow with Enosis. They would have preferred to achieve their aim by peaceful means, if possible, but they were never given the chance. Largely because of Turkey's implacable opposition, Britain never felt able to accept Enosis, even in principle. Archbishop Makarios was never prepared, until the very end, to consider anything less, and there was therefore little hope of a peaceful negotiated solution as long as Enosis remained the only issue.

The choice before the Ethnarch was therefore either to abandon his demand for Enosis or to resort to violence in order to achieve it; he chose the latter, and called on Grivas to fight the military campaign for him. The Greek-Cypriots continued to give him and the insurgents their support once violence started, even though it meant that they were now themselves directly involved.

They were indeed in a difficult position. The Church, the Greek leaders, EOKA and Athens Radio all preached incessantly that it was the duty of all patriotic Cypriots to fight for the liberty of their country and for its rightful union with Greece. Those who did not join the national struggle were publicly condemned as unworthy, cowardly and disloyal; hatred towards the British was patriotic and support of Enosis was a duty.

Such exhortations stirred many hearts, and EOKA did not lack recruits; above all, the younger generation responded, as will

[1] *The Times*, 17th December 1954.

youth anywhere, to a stirring call to arms for a national cause. Grivas was a firm believer in the use that could be made of youth movements, and he deliberately set out to arouse and to organize the pupils of the Cypriot schools to his own ends.[1] They responded with reckless enthusiasm and contributed substantially to the difficulties of the Security Forces; it proved a constant problem to control them, and also to find methods of punishing youths who had been involved in terrorism, but without exposing the Government to damaging propaganda charges of brutality and inhumanity to children.

Alongside the exhortations, there were also threats and brutal murders as part of the deliberate policy of formidable intimidation. Altogether during the Emergency EOKA killed more Greek-Cypriots than British and Turks together.[2] Citizens were constantly being threatened as 'traitors', and all too often they were later murdered in cold blood, sometimes in full view of their family and others, to make quite sure that the lesson went home. Some victims were beaten to death, one was burned alive and others were blown up by time-bombs. All went in fear of informers, schoolchildren spied on their teachers and none were safe.

The Government believed that the majority of the Greek-Cypriots did not approve of the terrorist methods used by EOKA, and that they supported EOKA only out of fear of the consequences if they did not. The Government aim therefore was to turn the people against the insurgents, and it was hoped that this might be possible. Various methods were tried, but, as we now know, they were none of them successful. This is not to say,

[1] 'The demonstration of Nicosia pupils which I ordered has been held . . . it was very good and militant.' (Letter by Grivas, 24th May 1955, quoted in *Terrorism in Cyprus*.)

[2] Total numbers killed by EOKA 1955–58 (by nationalities) were:

British	142
Greek	218
Turkish	29
Others	4
Total	393

however, that they were not fully justified at the time in the light of the evidence then available.

As a first approach, a determined effort was made to find an alternative to Enosis and thus to undermine EOKA, who persistently maintained that nothing short of Enosis was to be considered. This did not succeed, because Enosis was indeed what the Greek-Cypriots wanted, even if they did not approve of EOKA methods of trying to achieve it. So long therefore as Greece, the Church and EOKA all insisted that Enosis could still be obtained, but only through violence, the people had little choice but to follow. Only when their leaders abandoned their hopes of Enosis, did the people of Cyprus follow their lead and accept a compromise.

The British Government hoped constantly to find a moderate element among the Greek-Cypriots, with whom they might be able to negotiate an alternative solution and round whom they might rally opposition to EOKA. This was one of the objectives in mind when Archbishop Makarios was deported, but it did not materialize. If any moderate element did exist, it dared not show itself publicly, to face the threats of EOKA and the denunciations of the Church and Radio Athens.

Another method tried was to offer rewards for the capture or death of leading insurgents, but although the rewards were substantial (£10,000 for Grivas himself and £5,000 for some of his lieutenants), they remained mostly unclaimed. It is of interest that the Commissioner in Nicosia did on one occasion try out the technique employed by General Templer at Tanjong Malim[1] to obtain information without compromising the informants. But this time it failed, and he received back only blank sheets of paper.

When the 'carrots' failed to produce results, the Government tried a touch of the 'stick', and resorted to limited punitive measures, in an attempt to convince the people of the folly and the cost of supporting the insurgents.

Faced with a collective refusal by the public to co-operate with the Security Forces against the insurgents, the Government imposed collective fines on towns where it was proved beyond doubt

[1] See p. 70.

that people were refusing to reveal useful information known to be in their possession. The town of Limassol was fined £35,000 in September 1956, and Famagusta, with a population of only 25,000, was fined £40,000. On two occasions, the occupants of houses and shops, which it was thought were being used by the insurgents, were evicted and the property either destroyed or requisitioned.

Several lengthy curfews were imposed on villages, where guerrillas were thought to be hiding. One on Milikuri lasted fifty-four days, because it was reliably reported that Grivas himself was there; in fact he had been, but had slipped out just as the cordon closed in on the village, narrowly escaping capture, though this was not realized at the time.[1] Another curfew lasted forty-five days and one of nine days in Nicosia caused considerable hardship to the population. These curfews did in fact have a military purpose, but they were inevitably regarded by the populace as being purely punitive, and were certainly loudly condemned as such by the Greek propaganda machine.

The policy of collective punishment did not succeed in turning the people away from EOKA, and it was widely criticized as doing more harm than good to the cause of Britain, in that it aroused bitterness among the uncommitted element of the populace, and alienated potential goodwill. It was a step that was undoubtedly taken very reluctantly, but it was justifiable in that it might have helped to defeat EOKA and so end their intimidation of the populace. In fact, it confirmed that the insurgents' hold over the people was so strong that no action by the Security Forces, except the defeat of EOKA, was likely to overcome the fear that the people had of insurgent reprisals against alleged 'traitors'.

All these attempts to gain the support of the Greek-Cypriots, or at least to convert them from support of the insurgents, failed, and for the same reason. The Security Forces were never able at any stage to protect the populace effectively against reprisals, and until they could, there was little incentive to anyone to risk their life in a cause which held no promise of Enosis.

'We are all EOKA,' declared Dr. Dervis, the Mayor of Nicosia,

[1] *Cyprus Guerrilla* by D. Alostos, p. 133–5.

and he spoke with some justification, as well as with an eye to his own safety.

Had the Government been able to crush EOKA, or even to capture Grivas, the fear of intimidation would have been removed and opinions might have changed. But Grivas miraculously survived all attempts to capture him, and took care to demonstrate his ability to strike almost where and when he chose, so that he remained a very live deterrent to any would-be collaborator.

Otto Heilbrunn in his book *Partisan Warfare*[1] puts forward three basic anti-guerrilla principles for the handling of the local population, and it is interesting to apply them to the Cyprus Emergency. The populace, he says, must first of all be protected against the guerrillas, and secondly, must be converted from support of the guerrillas. We have seen that neither of these aims could be achieved in Cyprus.

The only course open, if these tactics fail, he states, is then to try to isolate the populace from the guerrillas, as was achieved in Malaya and Kenya by resettlement, food denial and the establishment of a Home Guard. Conditions in Cyprus, however, made it impossible to implement any of these measures, or to find any other means of isolating the Greek-Cypriots either mentally or physically from the insurgents and their incessant propaganda and ruthless intimidation. None of Otto Heilbrunn's three requirements could therefore be met.

The battle to win the hearts of the people of Cyprus was fought against considerable odds, and the sad fact is that, despite determined efforts, the Government did not succeed at any stage in winning the support of the local population. It is not the intention here to try to determine how much this lack of co-operation was due to genuine patriotism and how much to fear of EOKA, or to decide whether the populace might conceivably have been won over by some other methods. It is of interest, however, to assess the effect on the Security Forces.

The lack of public support was very much to the disadvantage of the Security Forces, who were driven as a result to operate throughout under a great handicap. They were faced with the

[1] Page 36.

vicious circle, which is common in varying degrees to all counter-insurgency campaigns. They could not defeat EOKA because of lack of information from the public; but until they defeated EOKA, they were unlikely to obtain this information, except by quite different means, based on their own efforts.

The insurgents, on the other hand, were able to operate under almost ideal conditions, assured of freedom of movement and secure in the knowledge that they could rely on obtaining food, shelter, information and protection at all times with virtually no risk of betrayal. These advantages were the fruits of their intimidation policy, and it made EOKA very hard to beat.

The Emergency in Cyprus proved again the decisive part played by the local population in any counter-insurgency campaign, and also the essential need for the Security Forces to be able to protect the populace if they are ever to win their support.

Public Relations

Archbishop Makarios realized early in the Cyprus crisis that he could gain formidable support for Enosis if he could rally public opinion outside Cyprus to his cause.[1] He succeeded in winning over the Greek Government, and then used them to arouse world opinion through the United Nations; he himself contributed significantly by his speeches and diplomatic activities.

He was helped by three factors. First, he could plead his country's right to 'self-determination', which was a cry arousing general sympathy. Secondly, Britain was vulnerable to attack as a 'colonial power', at a time when 'colonialism' was regarded as a sin, particularly after the Suez crisis. Finally, he had some limited Communist support, not for Enosis, but for any action he took which might be exploited by Communism as an embarrassment to the West and to NATO.

His case was presented to the world by a friendly Press in both Cyprus and Greece, by the propaganda of Radio Athens, and by

[1] The opening sentence of Grivas's Preparatory General Plan read: 'The Objective (is) to arouse international public opinion.' (*Grivas Memoirs*, p. 204.)

the diplomatic actions of Greece. Britain's reply took two forms; firstly, the negative act of curbing the hostile propaganda, and secondly, the constructive presentation of Britain's side of the case.

Full powers to control the Press were available under the Emergency Legislation from the start, but were very sparingly used. Radio Athens was not jammed until May 1956, more than twelve months after it had opened up. The Press in Cyprus was not suppressed, except during the Suez crisis, although the powers to do so existed under the Emergency Regulations. A 'stop list' was used to control the publication of news rather than a censorship, and there was only one case of a prosecution.[1]

The sophisticated and effective anti-British propaganda from Radio Athens and other sources demanded more than negative suppressive counter-measures to be taken against it, and the Government took energetic steps from the start to build up an efficient Public Relations and Information Service, which they then used to fight the all-important battle for public opinion.

A Director General of Information Services was appointed with direct access to the Governor, and a Central News Room was set up, where close liaison was established between civil, military and police information services. Emphasis was placed on the need to gain the confidence of the Press, and to ensure that there was a quick denial of falsehoods and a retort to any accusations or denigrations; this was important, not only to prevent an unfavourable impression being created abroad, but also to protect the Security Forces from unfair attacks. It was found also that it was very important to give a public explanation and justification of unpopular security measures, such as curfews, restrictions on movement and new legislation, because they would otherwise be quickly misinterpreted and used as propaganda by the opposition.

Efforts were made at the same time to develop a propaganda policy directed against the insurgents and their followers. It was based on stressing the folly and futility of insurgency, and tried to arouse support for possible political solutions other than Enosis;

[1] Charles Foley, editor of *The Times of Cyprus*, was tried for publishing 'statements likely to cause alarm and despondency' and was fined £50.

but it met with very little success, so long as Greece and Archbishop Makarios held out hopes of Enosis.

A third audience to be considered were the Security Forces themselves. They had a thankless task, abused constantly by their foes, which was to be expected, but also abused on occasions by their own Press and public figures, which they did not expect. Despite a robust defence by the Government in Cyprus and at home, a sense of distrust and resentment between them and the Press was difficult to avoid at times, and special efforts had to be made to emphasize to the troops and their commanders the importance of being 'Public Relations conscious'.

Public Relations was more important in Cyprus than in the other two Emergencies, because of the international aspect of the campaign, and also because propaganda was being so deliberately and effectively used as a weapon by the supporters of Enosis. One lesson from this campaign is that there is, in any cold war campaign, a real need to be aware of the Public Relations and propaganda issue from the start, and to provide not only a clear, firm Public Relations policy but also the machinery and the skilled, professional staff to implement it.

Civil-Military Relationship

The Cyprus Emergency demonstrated yet again the vital need for the closest possible civil-military co-operation and understanding. The Cyprus question was essentially a political issue, in which the military forces were acting in support of a political policy. Every military action had political implications, and there could seldom be any major decision that could be called 'a purely military matter'.

Furthermore, the conflict between short-term and long-term interests was particularly acute; the military measures required to achieve the short-term aim of defeating EOKA tended always to run counter to the longer-term Government interests of achieving a satisfactory political solution based on the goodwill of the populace. Measures that might seem essential on military grounds to fight the insurgents and to give the troops reasonable protection in the execution of their duty, were often opposed or modified on grounds of political expediency. This engendered a feeling of frustration and a sense of being required to fight a ruthless, treacherous enemy, 'with one hand tied behind the back'.

This problem was recognized by the Field-Marshal from the moment he took over, and, as a result, an excellent civil-military relationship was established from the start. As we have seen, he reorganized the command and control structure on a joint basis to include, not only all three Services, but also the Police and the Civil Administration. In particular, he appointed the Under Secretary (Internal Security), largely to ensure effective liaison. He integrated the Intelligence and the Information Services, and made it clear that all elements of the Government were fighting the same campaign against the same enemy and would therefore work to the same plan.

Relations between the Army and the Police were made much closer, and this liaison was put into practice at a Joint Army Police Staff School set up in November 1955, which was replaced early in 1957 by the 'Internal Security Training Centre'. These schools did excellent work in developing a joint approach to the campaign, and in training both military and police officers in the operational and staff techniques of counter-insurgency warfare; they also provided a centre for the pooling of ideas and the development of new methods for what was largely a new form of warfare, experienced before by the British Army only in Palestine from 1945 to 1948.

Another field in which mutual confidence was essential concerned the constant accusations levelled against the Security Forces by EOKA, Greece and even by some Britons of 'torture' and 'brutalities', particularly during interrogations. As in Kenya, there were isolated proven cases, but, on every occasion, firm disciplinary action was taken. The majority of the accusations were either exaggerated or false, and publicized for propaganda purposes. The Security Forces did, however, need to know that they would be championed and defended officially when they were falsely accused. They were constantly exposed to tremendous provocation, and only the highest standards of discipline and restraint prevented undue roughness or any retaliation. At the same time, they were subjected to a ceaseless propaganda campaign by a former wartime ally, in which they were not only reviled and falsely accused, but in which every action was distorted and exaggerated and then broadcast to the world.

The Security Forces themselves could not adequately reply to the charges against them, and had to rely largely on the Cyprus Government to defend them. They had a stalwart champion in the Field Marshal, who fully appreciated how much this meant to the morale of the troops and the police, and to their confidence in their political leaders. He gave them the dictum of 'Courtesy with Firmness' and enforced it; he tolerated absolutely no indiscipline, but at the same time, he fully supported any justifiable action, and took all possible steps to protect the troops against their accusers and against legal action for acts committed on duty.

The mutual confidence that this built up greatly helped and encouraged the Security Forces in their most difficult task; and they in their turn took pains not to abuse their powers. Nevertheless, the Government attitude inevitably made them vulnerable to criticism and to charges of condoning brutality, repression and the denial of human rights and liberties.

On the whole, the Cyprus Emergency provides an example of a close and effective civil-military relationship, which was subjected to considerable strain but stood the test well, and is a good model for future situations on similar lines.

Summary

The Cyprus Emergency provided the most formidable opposition, both political and military, that Britain had faced so far, except possibly in Palestine 1945–48. Indeed, there are several interesting similarities between the two campaigns. In both cases, for instance, the Security Forces fought under the great handicap of facing a persistently hostile populace; and in Cyprus, as in Palestine, the discipline and restraint of the troops were outstanding, despite intense provocation. There can be few, if any, other

Armies, which could have maintained equal standards under the circumstances.

It was a campaign fought in highly developed, European terrain, and with great emphasis on the political as well as the military struggle. The military outcome was a somewhat negative one in that EOKA were not decisively defeated; but they were nevertheless prevented from winning, which was equally important. It was largely thanks to the efforts of the Security Forces that it was possible to achieve a political solution which was rather more favourable to Britain than to EOKA.

The Emergency in Cyprus is a classic example of a campaign against skilled and sophisticated insurgents, fighting with strong political support, and in which the behaviour of the local population was of great significance. The campaign is worthy of detailed study, for it offers many lessons of value for future counter-insurgency campaigning.

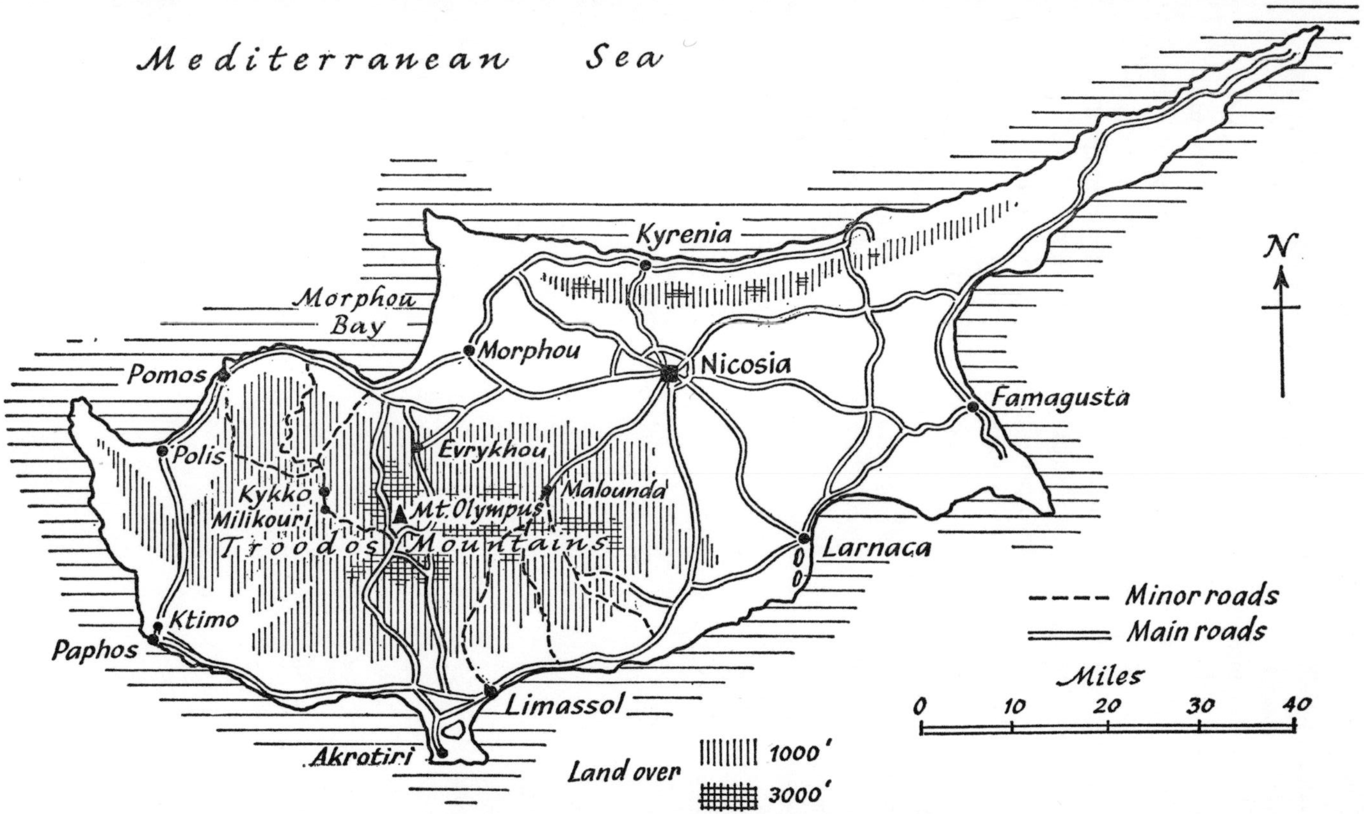

Map III. CYPRUS, 1956

Chapter 16

FIGHTING THE INSURGENTS

'We should realise that unorthodox kinds of warfare are bound to be employed on a very wide scale in any future war, and the matter should receive the special attention of the staffs of every country.'

This dictum was laid down not by the Chiefs of Staff of Britain or the United States, but is the view of a well-known and experienced insurgent leader called General Grivas.[1] He wrote this in 1964 and it is very sound advice.

It is a reasonable assumption that the British Army is likely to be engaged over the next twenty years in very much the same type of counter-insurgency operations as during the last twenty years. There will be some changes, of course; the origins and aims of the insurgencies will vary and it cannot be foreseen with any certainty who the enemy will be. British counter-insurgency forces of the future may well be part of an international or even a United Nations Force, and they are increasingly likely to find themselves fighting on foreign soil, such as Borneo and Vietnam. But the strategy and the tactics of insurgency will not change greatly, even though the political implications are unpredictable and the weapons of the insurgents may become increasingly sophisticated. It is important therefore that the lessons of the past should not be forgotten or ignored, but should be applied to the future.

There is much to be learned from the three Emergencies which have just been reviewed, and from the many similar campaigns

[1] *Guerrilla Warfare* by General Grivas, p. xii.

which have also been fought by other countries. This chapter aims to consolidate this experience into an analysis of the main essentials and principles of counter-insurgency warfare, as it might be applied to present and future campaigns. It falls into four parts. First, a definition of the aim; secondly, a study of the essential requirements for successful counter-insurgency operations; thirdly, possible methods to be adopted against insurgents, and finally, some thoughts on how the vital 'battle for the hearts and minds' of the people can be fought and won.

THE AIM

The first essential in any counter-insurgency campaign is that both the political and the military aims should be agreed by all concerned from the very start and should be clearly stated in a directive. As the military aim is dependent on political considerations, it is essential that the political directive should clearly state three points; firstly, the purpose and scope of military operations; secondly, the short-term political and military aims governing the campaign, and finally, the long-term political aim which it is hoped to achieve when the military campaign is over.

The military aim is normally to defeat or to neutralize the insurgents as soon as possible, so that lawful government can be restored and a political settlement can be reached, based on the goodwill and the free wishes of the populace. Speed is an essential part of this aim, so that the conflict can be limited or quelled before it can develop into a threat to world peace. This calls for strategic mobility in order to achieve effective intervention without delay; one battalion quickly on the scene may well prevent a protracted and costly campaign by many brigades.

THE ESSENTIALS FOR COUNTER-INSURGENCY OPERATIONS

All counter-insurgency forces have certain essential requirements, in the same way that the insurgents do, and they are:

(*a*) Civil-military understanding.
(*b*) A joint command and control structure.
(*c*) Good intelligence.
(*d*) Mobility.
(*e*) Training.

These requirements are different from those in conventional warfare in several ways, and it is not sufficient merely to adapt conventional warfare methods to meet the special conditions of counter-insurgency campaigning.

Civil-Military Understanding

It is to be hoped that the days of mutual mistrust and misunderstanding between soldiers and civilians are now past, and that both now appreciate the need to work together as one team.[1] Nevertheless, one of the outstanding lessons of all cold war campaigns to date has been how vital this understanding is in this particular type of warfare, and at all levels from the Cabinet to the Company Commander.

As General Templer constantly emphasized in Malaya, Civil Government and military operations are part of one plan against a common enemy, and they cannot be kept apart.

There is no purely military battlefield in counter-insurgency warfare, and the campaign can only be won by combined civil-military efforts. Many military measures cannot be effective without the support of the Civil Administration, and conversely, wise government can often ease the task of the Security Forces or even finish it for them. In both Malaya and Kenya, for example, the establishment of a Home Guard was a most decisive step, and the same was true of the resettlement programmes. In the Huk insurgency in the Philippines the revolt was stifled by the imaginative policy of the Defence Minister, Ramon Magsaysay, in offering such generous surrender terms that the majority of the insurgents lost the will to fight on.

[1] 'Military and political problems are not separable, and military and political men must work even more closely together.' (President Kennedy in a speech to the NATO Military Staff on 10th April 1961.)

There are equally points on which civil and military are liable to disagree, due to the differing viewpoints that they each tend to have of the situation. The military commander is primarily concerned to defeat the enemy as soon as possible, and at the same time to protect the lives and maintain the morale of his troops. As a result, he will at times feel that his freedom of action is being unduly restricted by political considerations.

The civil authorities, for their part, have their eyes primarily on their long-term settlement, based on the goodwill of the people, and they are liable to feel sometimes that they are being pressured by unreasonable military demands into action which is prejudicial to this aim. It is a matter always of priorities, and these priorities will themselves change constantly as the situation changes.

A really sound civil-military liaison and understanding is essential under these circumstances, and there are three main ways in which it can be achieved. Firstly, as already mentioned, the Government must, at an early stage, declare its national policy, and at the same time define the military role in that policy, so that all concerned know and can work to one common purpose and plan.

The second condition which is essential to effective civil-military co-operation is the establishment of a unified command, with the direction and control of the campaign in the hands of one person. Only in this way can complete mutual confidence between civil, military, police and Whitehall be ensured and a firm control of the campaign be maintained. The truth of this principle was proved by the appointments of Field Marshal Harding and General Templer, but it is not always possible to find the man to fit the post.

Thirdly, it is of great value if the civil and military organizations can work, plan and prepare together as much as possible *before* the crisis arises, rather than having to build up their teamwork after the trouble has begun. There must be the closest possible co-ordination between the Armed Forces, the Police, the Government Departments and the intelligence organizations, both civil and military, not only in the territories where insurgency is likely, but also in Whitehall. Cold war campaigning and counter-insurgency both need to be studied, planned and controlled by an integrated team, in peacetime as well as when active operations start.

Above all, it is essential to have effective machinery to ensure full consultation and co-ordination between civil and military during the campaign and to enable the decisions taken to be swiftly and effectively implemented.

Command and Control

In any counter-insurgency campaign, and particularly in a colonial territory, a special command and control structure is required, in view of the many authorities involved and the problems of proper co-ordination. Experience has shown that there are four essentials, which are:

(*a*) A Director of Operations.
(*b*) A joint committee structure.
(*c*) A Secretariat.
(*d*) Joint Operations Rooms.

The requirement for one overall *Director of Operations* has been demonstrated on every counter-insurgency campaign to date. These operations involve joint action and planning by so many different interests that effective co-ordination can only be achieved if complete responsibility is in the hands of one man. He must either be the Head of Government himself, or be directly responsible to him; but in either case, he must be in a position to exercise direct control over all security operations.

This arrangement does not alter the principle that military operations are always subordinate to political considerations; but it avoids the unsatisfactory situation where the Army acts 'in support of the Police', as occurred in the early stages of the Emergency in Malaya, and in Aden until 1965. Such a relationship may be adequate for Internal Security or Policing operations, but it is quite unsuited to all-out counter-insurgency warfare, where a military commander is required in full control, with the Police and other elements in support of the Army.

The Director of Operations should be served by a fully integrated staff consisting of civil, military, Police and Intelligence personnel, working together as one team. This organization should continue downwards throughout the normal 'chain of command' and there should be a joint staff with a Joint Operations Centre at every level involved in the planning and direction of operations.

Alongside this executive staff there is a requirement for a joint structure of committees to plan and control operations. This does not mean 'command by committee', which is to be avoided at all costs; it does ensure, however, that there is proper 'discussion and decision' between the many authorities concerned in the planning and implementation of operations.

The *joint committee structure* should be based on the long-established triumvirate of 'civil-military-police', which has been the backbone of many counter-insurgency and Internal Security campaigns in the past. The Intelligence aspect has not always received the weight due to it, and it should also be included in the team as a fourth member. Other bodies such as Public Relations, Psychological Warfare, Legal Advisers and Security Advisers have a part to play, but need be called in only when specially required.

The details and the nomenclature of the committee structure will vary, but the basic pattern is clear. At the top is a Policy Committee, which is responsible for the general conduct of the campaign; it will be headed by the supreme authority, be he civil or military, with the Director of Operations, the Chief of Intelligence and the Commissioner of Police as members. Others, who will be called in as required, will be Ministers and Advisers, but this body is in the nature of a Cabinet or War Council, and must be kept small, with the emphasis on policy decisions rather than detailed planning.

The decisions of the Policy Committee will be implemented and the day-to-day conduct of the campaign will be controlled by an Operational Committee. This body will be headed by the Director of Operations, with the usual civil, Police and Intelligence representatives as members. The civil representative will often be a specially created post, as for example the Secretary (Internal Security) in Cyprus and the Defence Secretary in Malaya.

The chief executive body, is the Operational Committee maintaining direct day-to-day control of all aspects of the campaign and issuing orders to implement the policy of the Government in both civil and military affairs. It will receive and assess all information, and will advise the Head of Government on all counter-insurgency matters.

Its decisions and orders will be passed downwards through a chain of similar bodies at each level running parallel to, but quite separate from, the military chain of command. In every State, Region or military Command, there will be a similar committee with the appropriate members, and it will receive and act on the orders and policies of the committee above it. Eventually, at the 'sharp end', it will be a case of the Company Commander planning his patrol programme with the Battalion Intelligence Officer, the local Police Post officer and a young District Officer.

This may seem a cumbersome system, but it does in fact ensure that the best possible use is made of all information, all resources and all ideas; it also makes possible the integration of the four elements of any counter-insurgency campaigning, which cannot operate in isolation without causing conflict and confusion; Malaya provided a good example of the structure in action.

To complete the structure for civil-military co-ordination there is a need for a *Secretariat*, to serve the top-level committees and to form one co-ordinating point for all the bodies concerned in the conduct of the campaign. Apart from the routine business of agendas, briefs and minutes, the Secretariat will be responsible for any 'follow-up action', and for overall co-ordination.

The Secretariat must itself be essentially civil-military in both organization and outlook; the staff should consist of military and civil officers, and possibly Police and Intelligence representatives. It should be responsible to the Head of Government, with bias in no direction, but insight in all directions.

Firm executive control of any military operations calls for an Operations Room, and in the case of counter-insurgency, there must be *Joint Operation Centres*, where the decisions of the appropriate Operational Committee are put into effect. They must also be jointly manned by the military and police; it is usually valuable to include the civil side as well, and failing this, the closest possible liaison must be maintained with the civil and intelligence representatives. They will be required at each level in the chain of command, must be completely secure, and have good and reliable communications.

Good Intelligence

It is not easy to discuss intelligence matters usefully in public, firstly, because there is inevitably some security risk, and secondly, because few outside the world of intelligence can be in a position to make any comments with full knowledge and authority. But the subject is so vital to any study of counter-insurgency techniques that it cannot be omitted.

Counter-insurgency intelligence must cover a wide field and deal not only with the operational organization and capabilities of the insurgents, but must try also to expose and to understand their minds, their mentality and their motives. The influence that they are likely to exert over the populace must also be anticipated so that their efforts at subversion and intimidation can be thwarted. This is comparatively easy in British colonial territories, (though it has not always been successful in the past), but it will be more difficult when British troops are called upon to operate in other countries.

The organization of intelligence during a counter-insurgency campaign follows three phases, corresponding roughly to those of the campaign itself, as outlined in Chapter 3. The first phase is that leading up to the outbreak of operations by the Security Forces; the Police Special Branch will usually have fairly good information at this stage, for the opposition will still be an overt organization. The problem is to take effective action against them, due to lack of adequate powers without the added strength of Emergency Regulations; also, political considerations usually make it difficult to adopt a firm line at this stage.

The second stage is reached when the insurgents go 'underground' and the Armed Forces begin operations against them in conjunction with the Police. There is then an urgent need for integration of all intelligence efforts and resources, civil, police and military, but new machinery to achieve this cannot be created easily or quickly. It should ideally be in existence at all times, so that it does not have to be built up hurriedly when operations start. One essential is to appoint one Chief of Intelligence to be responsible to the Head of Government for all intelligence matters, and so be in a position to draw all the strings together. Another requirement is for skilled interrogators, since this may often be the main source of intelligence.

Intelligence will not be easy to acquire at this stage, due to lack of co-operation by the populace, and to lack of good sources of information. The Security Forces must therefore make the gaining of intelligence one of their top priority tasks; patrols, observation and, above all, prisoners who can be interrogated are of the greatest value. It may be a long and tedious search before the breakthrough occurs, but it always does come sooner or later,[1] and intelligence then should improve considerably. Cyprus is a good example of what can be achieved in the way of gaining intelligence without any co-operation from the local population, and it proves that success is not entirely dependent on their support.

The final stage is reached either when the populace decide to co-operate and to give intelligence against the insurgents or else when the Security Forces have gained enough information from their own sources to be able to penetrate and so break up the insurgent organization. This moment is like reaching the top of a hill, for once the enemy are evidently on the run, information usually snowballs, and it is a much easier journey down the other side.

Good intelligence is undoubtedly one of the greatest battle-

[1] The search of Nairobi (p. 98) revealed the Mau Mau order of battle, while in Aden the unexpected capture of one terrorist in 1965 provided many details of the terrorist organization there, and led to several successful operations against them.

winning factors in counter-insurgency warfare, and the acquisition of it depends on the following six principles:

(*a*) Money and effort spent on the gaining of intelligence and the preparation of an effective organization *before* the insurgency begins will pay handsome dividends in the long run.[1]
(*b*) Every effort must be made to 'Know the Enemy' as thoroughly as possible before the insurgency starts.
(*c*) The best results are achieved from a fully integrated intelligence organization, under one Chief of Intelligence.
(*d*) Intelligence must be worked for and not waited for, and the acquisition of it must be made a top priority task for the Security Forces, particularly in the early stages of a campaign.
(*e*) The co-operation of the populace is not essential to the gaining of intelligence, but it is a tremendous asset, and every effort must be made to win this support.
(*f*) The best intelligence comes from penetrating the insurgent organization.

Mobility

In order to tackle any insurgency early and prevent it spreading, there is a need for strategic mobility, so that the appropriate forces can be sent to the scene without delay and can go into action immediately on arrival. To do this they must be fully airportable and have the necessary aircraft available, as is the case with the British Strategic Reserve.

On arrival, they need local mobility in the form of good communications by land and air, which means suitable roads, transport, stores and airfields. The helicopter is outstandingly valuable because of its versatility and the mobility that it confers on the

[1] 'If we had had (in Malaya) in 1948 the Police Special Branch system that we had built up by 1954, the insurgency would never have gotten into it's stride and would certainly have been ended more quickly.' (*Communist Defeat in Malaya* by Colonel R. L. Clutterbuck, March 1954 in the Royal Artillery Journal.)

fighting units, as was shown in the later stages of the Cyprus Emergency.[1] Air supply gives further mobility, but is only of maximum value if the troops are sufficiently highly trained to carry out the type of operations calling for it.

One lesson of every Emergency to date has been the value of mobile and aggressive patrolling, and the ability to penetrate deep into the guerrilla territory and then employ guerrilla tactics against them. This cannot be achieved simply by providing the right equipment, air support or any other aids. Tactical mobility depends in the end on two factors, training and morale. No troops can operate efficiently in undeveloped country in small numbers without experience and the confidence that springs from it, and these cannot be acquired overnight. Ultimately the will to master the terrain and the enemy is the decisive factor which gives the infantryman his mobility in action. This precious quality can only be acquired by hard training; it cannot be issued like a piece of equipment. But once the Security Forces can match the insurgents at mobility, they are well on the way to mastering them completely.

Training

Since 1945, the British Army has spent some seven times as long engaged in actual counter-insurgency operations as it has on conventional warfare campaigns, and this seems likely to continue. This ratio has not, however, been reflected in our training, where, until very recently, greater emphasis was placed on training both staff and units for nuclear or conventional warfare rather than for the type of warfare which we have been actually fighting almost continuously for twenty years.

Counter-insurgency warfare is a highly specialized subject, calling for continual study and practice, and to be fully effective, training for it must be carried out jointly by all those directly concerned, both civil and military; this applies to planning and

[1] In Malaysia today, it is calculated that the ratio of time for movement in the jungle by helicopter as opposed to by foot is about 1 to 60, i.e. 10 minutes by helicopter equals 10 hours on foot.

staff work just as much as to operational techniques. It is not a subject that can be properly studied or practised by any part of the Security Forces in isolation, and this must be the first lesson to be taught.

It is reasonably simple to find a basis for conventional warfare training, in that we have a very fair idea of the type of enemy we shall be fighting, the weapons and tactics that he will use, and we know that, within limits, certain rules are likely to be followed. This is far less so in counter-insurgency warfare, where the circumstances are highly unpredictable; the enemy will range from Mau Mau to EOKA, and in the future from Communist 'volunteers' perhaps to 'insurgents' worldwide in a variety of forms. Each has very different characteristics and calls for different techniques by the Security Forces; the terrain too cannot be foreseen and will vary from jungle and swamps to mountains or main streets.

Training must include a thorough knowledge of the probable enemy, his methods and his weaknesses, as well as a study of the civil aspects of cold war campaigning, so that all ranks understand the problems of the civil authorities and appreciate their aims.

Training in counter-insurgency can prepare troops for most of the likely situations, but it is a far cry from being skilled in 'APC drills' in Germany to carrying out 'ambush drills' in the jungles of Borneo; it inevitably takes some time for even the best troops to adapt themselves from one specialized technique to another. The quelling of an insurgency calls for quick action by troops trained for this task, for there is no time to learn the techniques once the campaign has begun, if the essential early success is to be achieved.

Counter-insurgency warfare calls for small-scale operations of section or platoon strength, where a handful of men is required to fight in complete isolation for long periods under conditions which demand the highest standards of initiative, resource and junior leadership. This is a skill which is in marked contrast to the complicated, large-scale battles of modern conventional warfare, and it is not easily or swiftly acquired. Yet it is the basis of all good infantry training, and is also the likeliest form of operation in the foreseeable future; it is also very appropriate to nuclear warfare.[1]

At the higher level, there is also a need for joint training in the planning and staff work involved in counter-insurgency warfare, together with its cold war implications. There would be great value in holding short courses at existing military schools and Staff Colleges for representatives not only from the Armed Forces, but also from the Police, the Intelligence agencies, and Government Departments. These courses should cover all aspects of counter-insurgency and should study the subject in its broadest aspects, in order to analyse the lessons and experience of the past, to exchange ideas and provide new thought. In overseas territories, where there is any likelihood of counter-insurgency operations developing, joint planning and joint training should certainly be carried out, particularly between the Army and the Police, with close air support.

There is even justification for a Cold War Study Centre, devoted purely to research into cold war campaigning in all its aspects, military, political, economic and ideological. It could devote more time and thought to this complex subject than can the planners who are battling with urgent everyday problems, and it could also consult all possible sources of knowledge and experience, civil and military, official and unofficial, national and international.

DEFEATING THE INSURGENTS

Insurgents cannot be defeated by conventional warfare techniques, and a completely fresh approach is required.

The most difficult task in counter-insurgency warfare is to make contact with the insurgents and to kill them in battle; a totally disproportionate effort is required for every insurgent eliminated in this way, and it is a lengthy and laborious process. It is far more effective, and also more economical, to defeat them by making it

[1] 'Stalking terrorists in the Malayan jungle is today, strange as it may seem, the best training for nuclear warfare. The use of new weapons and technical devices can quickly be taught; to develop hardihood, initiative, mutual confidence and stark leadership takes longer.' (*Defeat into Victory* by Field-Marshal Slim.)

impossible for them to fight on; this can be done by depriving them of those essentials on which they depend to survive. To pursue Mao Tse-tung's simile of the insurgents living among the populace of a country 'like fish in water', the policy must be to destroy the fish either by removing the oxygen from the water or by draining the pond, rather than by chasing individual fish in the muddy and weedy water with a very small net.

An insurgency force, wherever it is fighting, cannot survive for long without certain essentials, which are:

(*a*) The support of the local population.
(*b*) Bases.
(*c*) Mobility.
(*d*) Supplies and information.
(*e*) The will to win.

How can the Security Forces deprive them of these essentials?

The Support of the Local Population

The local population of any country suffering an insurgency will normally fall into four categories. First, there are the active, willing supporters of the insurgents, and they can only be treated in much the same way as the insurgents themselves. At the other end of the scale are the active loyalist supporters of the Government, who must be protected and encouraged as far as possible. In the middle are the two vital categories, who hold the key to the outcome of the campaign. They are firstly, those who give their support to the insurgents but only out of fear of the consequences if they do not; and secondly, those who are not subject to intimidation, but who nevertheless sit firmly on the fence, waiting to see which way it will be best for them to jump in due course.

The aim of the Government must be to persuade or to force the last two categories firstly, to cease supporting the insurgents, and secondly, to transfer their allegiance to the Government. The methods to be used are based on the 'carrot and stick' technique, and they must be used with considerable skill and subtlety, if they are to be successful. Here we consider only how to deny the support of the populace to the insurgents; the winning of local support for the Government cause is discussed later.

One powerful 'stick' in the battle for local support is the use of reprisals and harsh punitive measures applied in the hope of making the populace more frightened of the Security Forces than they are of the insurgents. This is not likely to be an acceptable policy to any Western Power today, for it is morally unacceptable that any modern Government should try to outdo the average insurgent force in terrorism. And even if it were attempted, it is unlikely that it would be successful, for the insurgents will usually have the last word.

There are, however, punitive measures which are less severe, but yet may help to dissuade the populace from supporting the insurgents; these are collective punishments, such as curfews, collective fines, detention of suspects, and various restrictions on individual liberties. Collective punishment as a regular policy is morally undesirable, but it can be justified on occasions, if there is clear evidence of collective failure to co-operate with the Government as required by the law, of all citizens. Collective punishments will only achieve their object, however, if those who do, as a result of them, collaborate with the authorities can be protected against reprisals by the insurgents for so doing. This is one reason why they never succeeded fully in Cyprus, but did in Malaya from 1952 onwards.

It is an essential first step therefore that the populace must be protected against intimidation, and the most effective method of achieving this is to isolate them physically from all contact with the insurgents. This not only enables the Security Forces to offer them greater security, but also makes it physically more difficult for the insurgents to dominate them and to use them for their own ends.

This isolation is achieved mainly by the three methods already discussed, that is to say, a policy of food denial, resettlement, and the establishment of a Home Guard. These techniques proved remarkably successful in both Malaya and Kenya, but they could not be implemented fully in the urban conditions existing in Cyprus.

Measures to prevent the insurgents from having any access to the local population are liable to be unpopular initially with the populace because they interfere with their liberties or their earnings, and the insurgent propaganda will undoubtedly exploit this to the full. It is important therefore that the measures taken are fully and carefully explained and justified to the public, and that the public are made to realize that the battle against the rebels is their responsibility as well as that of the Security Forces, and that it is their duty, as well as being in their interests, to play their part.

Bases

Guerrillas need a base somewhere from which to operate, and from where they can be organized and controlled. It will usually be in a highly inaccessible area, difficult to locate or to attack; it may be in the theatre of operations, or, on occasions, it may be outside the theatre of operations altogether, as in the Korean and Vietnamese wars.

Just as a diver attacked by an octopus (to continue our fishy similes) will attack the beast's head rather than try to cut off each tentacle, so the counter-insurgent forces should make the enemy bases their first priority target in preference to pursuing individual gangs. This was successfully done in Kenya when Nairobi was cordoned off and thoroughly searched; the Mau Mau Headquarters was located there, and as a result, it was completely neutralized and never regained control over its forest gangs again. In Cyprus, well-planned operations were mounted against the EOKA hard-core gangs in the Troodos Mountains, and only just failed to achieve complete success by capturing Grivas. In Malaya, it proved virtually impossible to pinpoint the guerrilla bases, despite determined and continued efforts, but action was taken against the political organization instead.

Operations against the insurgent bases depend on good intelligence if they are to be effective, and they are therefore more likely to be successful in the later stages of a campaign. They may consist of large-scale operations designed to surround the base and eliminate all within the cordon, such as the searches in the Troodos

Mountains already mentioned; or they may be small deep-penetration patrols, which set out to pinpoint the base and then to hunt and harass the occupants, and in particular, the leaders. The latter tactics are usually more effective, and more feared by the insurgents, but they do require highly skilled and knowledgeable troops.

The air forces have a considerable part to play in any action against enemy bases, and helicopters are of particular value; they can provide reconnaissance and photographs in the preparatory stages, and can then help greatly to achieve surprise by dropping troops in position to cut off or surround the enemy. They then carry out air supply and evacuation of casualties.

Where insurgent bases are deliberately set up astride an international border, special political action will be needed, and the police will be called on, as well as troops, to control the border. The Royal Navy may also be required to carry out anti-smuggling patrols, where the insurgent forces are being supplied from an overseas base as was the case in Cyprus.

Mobility

A basic principle for guerrilla forces is to avoid set-piece battles, and to rely on harassing, hit-and-run tactics. It is greatly to the advantage of the counter-insurgent forces, therefore, if they can deprive the enemy of his freedom of movement; in this way, they remove his ability to strike where and when he likes, and also they gain the initiative.

To achieve this requires a combination of defensive and offensive measures. Static guards must be avoided as far as possible, though standing patrols and OP's are different; they can be of value, both by providing information and also by making the enemy realize that he is being watched. Road blocks, patrols and ambushes also serve to make the guerrillas more cautious in their movements,[1] but they do not completely deprive them of their mobility; this can best be achieved by three lines of positive action.

[1] 'The system of counter-ambushes [in Cyprus] was excellent . . . it was the only one of which we took some account. We always stood in fear of the "lurking cat".' (*Guerilla Warfare* by General Grivas, p. 52.)

Firstly, mobile reserves must be available to follow up the insurgents whenever contact is made. Not only must the enemy be denied the opportunity to escape at leisure, but they should be relentlessly pursued as far and as fast as possible. Having established a contact by knowing where the attack took place, the Security Forces must take advantage of it to gain the initiative and harry the enemy. This method was used with success in Malaya, and led to the discovery and so to the destruction of many Communist camps and gangs. It was also employed in Cyprus, where tracker dogs were used to find the terrorist hiding-places, while helicopters enabled the troops to watch the countryside and so prevent or intercept insurgent movement to an extent which would have been impossible from the ground.

Isolating the populace from the insurgents succeeded in Malaya and Kenya in denying mobility to the insurgents, because it deprived them of intelligence, food and shelter. Without intelligence about the Security Forces, the insurgents have to move more cautiously for fear of attack or ambush; without easy access to food, they have to move farther afield and to known areas to obtain their own supplies. Without the shelter that the villages offered, they have to stay in the jungle or forest.

Registration of the population is another means of checking the movement of the insurgents, and its effectiveness can be judged by the efforts of the Communists in Malaya to prevent it being implemented.

The most effective method of all, of course, is to win the support of the populace for the Government, for once they co-operate, it is extremely difficult for the insurgents to move anywhere without it becoming known to the Security Forces.

Supplies and Information

We have seen how insurgents depend on the local population for their supplies and their information, and how both these essential needs can be denied to them by a policy of isolation, which includes food denial.

The insurgents can usually rely on obtaining ample information about the Security Forces, because of their network of spies and informers carefully set up in the early stages of their bid for power. Their intelligence is therefore excellent, and any measures to prevent the leakage of information can pay good dividends in depriving the insurgents of this advantage. The alternative is, of course, to resort to deception and pass false information, as has been done with success many times.

The Will to Win

In an insurgent campaign, each side is constantly trying to undermine and to destroy its opponent's will to fight on, for success in that direction means victory, and it can be one of the easiest and cheapest methods of winning the war.

The morale and will to win of any insurgent force depend very much on having a stirring and convincing cause for which to fight, and also a belief that they are on the winning side. There are several ways in which their will to win can be weakened, the most effective of which is to neutralize the cause for which they are fighting. This cannot always be achieved, but it was in Cyprus, when Archbishop Makarios was forced to abandon his demand for Enosis and to accept independence. EOKA were then faced with the choice between fighting on alone and against his orders, or giving up the struggle; happily they chose the latter. In Malaya, the Communists used as the cause for which they were raising the banner of revolt the promise of an 'independent' Malaya freed from British 'oppression'; but this lost most of its power when Britain forestalled them by voluntarily granting Malaya her independence.

Propaganda and psychological warfare properly applied can also be used effectively to undermine the insurgents' belief in their cause, by sowing seeds of doubt and suspicion in their minds and by countering some of the indoctrination from which all insurgents suffer. It is important in all propaganda that the enemy should always be able to confirm for himself the accuracy of at least some of the information that is fed to him, so that he can be convinced of its truth. For example, references can be made to insurgents who have been killed or who have surrendered, and to actions of which it is likely that the insurgents in question will have personal knowledge.

This weapon can also be used to persuade insurgents to surrender, chiefly by portraying the danger and the hopelessness of their position as opposed to the safety and comparative comfort that awaits them if they give themselves up. Alongside the appeals there must also be convincing incentives, for the insurgent will want to be reasonably sure of the fate that awaits him before he makes his decision to desert. The lenient treatment of prisoners usually encourages defection, for news of this will reach the surviving insurgents somehow; one way of making sure that they know is to arrange for surrendered insurgents to sponsor leaflets or to broadcast by voice aircraft. In Malaya the Communist guerrillas used to ask villagers whether the life of Surrendered Enemy Personnel, as they were called, was really as rosy as the psychological warfare leaflets made out.

Generous surrender terms can also pay worthwhile dividends, but they need to be very carefully timed and presented; they should avoid any suggestion that the insurgents have necessarily been 'defeated', but rather that they are misguidedly supporting a lost cause, and that if they abandon it, they can return eventually to a normal life.[1]

Insurgent morale will usually be undermined and shaken if the Security Forces can inflict several decisive defeats on the enemy; but this cannot unfortunately always be arranged, and even if it is achieved, the effects are liable to be countered by insurgent propaganda.

The Initiative

The task of the insurgents is made far easier if they are allowed to retain the initiative and to attack on their own terms. The Security Forces must therefore make every effort to move from the defensive as soon as possible and to gain the initiative for themselves. This can be achieved by:

[1] Cf. the reaction of EOKA to the surrender call of 1956 (p. 134).

(*a*) Offensive action.
(*b*) Expansion from a firm base.
(*c*) Mobility.

Offensive action need not be on a large scale, and may consist only of active and aggressive patrolling from a firm base, in order to dominate the areas required by the insurgents. This will usually mean placing a military unit in a town, village or settlement, thereby reinforcing the police and also encouraging the populace by its presence. Patrols, road checks and ambushes must then be established in the surrounding countryside, so as to form a 'cordon sanitaire' round the inhabited area and enable the process of isolation to begin.

The firm base thus established should be expanded gradually and systematically, so that the insurgents are excluded from complete areas one after another. This forces them to fight the Security Forces more on terms chosen by the latter, and begins to deprive them of the initiative. It is a mistake to disperse the Security Forces on scattered operations over a wide area, unless they are strong enough to dominate that area; if they are weak, they expose themselves as ideal targets for guerrilla attacks.

As each area is freed of insurgent domination, it should be handed over to the local Armed Forces or Home Guard for protection and to the civil authorities for administration. This not only frees trained troops for more offensive operations, but also gives the Government an opportunity to win the support of the populace by wise and beneficent government.

Once the Security Forces are released from local operations around the populated areas, they can set about pursuing the insurgents farther afield, following them up relentlessly wherever contact is made and harassing them at every opportunity. This calls for mobility to enable them to outmarch and outmanœuvre the enemy, and can be achieved in several ways. The troops must be lightly equipped and suitably organized for small-scale operations, and must be fully airportable; the air forces can contribute tremendously to the mobility of the ground troops, particularly by the use of helicopters, which did more than any other measure to deprive the insurgents of their mobility in Cyprus, and in Malaya; in Vietnam they are the basis of almost all tactical mobility. Air supply also plays an important role.

These methods are more easily applied in undeveloped terrain such as Malaya and Kenya, than in the less primitive conditions of Cyprus or Aden town, but it will always pay to try to clear the enemy systematically from certain areas, and to confine him to others rather than to try to fight him with equal effort on all fronts. The more the insurgent organization can be broken up and each part separated from the others, the more difficult it becomes for them to retain the overall initiative.

WINNING HEARTS AND MINDS

We have seen that the support of the local population is an important factor for both sides in any counter-insurgency campaign, but particularly for the insurgents, who will resort to ruthless intimidation, if necessary, in order to achieve it.

Short-term local support is not the same, however, as the long-term objective of winning the hearts and minds of the people permanently. This is the ultimate aim of both sides, and is a predominant factor in all their thinking.

There are two basic requirements to be met before the support of the local population can be won by the counter-insurgent forces, either in the short or the long term. Firstly, the Government must demonstrate its determination and its ability to defeat the insurgents,[1] for no one likes backing a loser, particularly in an insurgency. Secondly, the Government must convince the populace that it can and will protect its supporters against the insurgents, for no one likes being shot as the reward for loyalty.

Given these two requirements, there are five main factors affecting the winning of the hearts and minds of the people to the Government and its cause. Most of them call for civil rather than military action, but the Security Forces can contribute substantially to their success or failure.

[1] 'Public opinion is ultimately gained by great victories' (Clausewitz).

The first essential is that those in control must understand and respect the feelings and aspirations of the nation. They must also take positive steps to prove this sympathy, so that the people are convinced that victory over the insurgents will bring them a political settlement which will meet their wishes. In Malaya, the promise of independence did just that, whereas in Cyprus, Britain could not win over the Greek-Cypriots, because she was not prepared to agree to Enosis. Similarly, the French refusal to grant Vietnam independence except within the French Union lost her the sympathy of the Vietnamese people and was one of the factors contributing to her ultimate defeat.

Secondly, there must be firm and fair Government, so that law and order are maintained and the interests of the populace are safeguarded, not exploited, by those in power. The Government may be either internal or external, but in either case, corruption and dissension must be eliminated, justice must be done (and be seen to be done!) and the administration of the country must be as efficient as it can be made in the face of the inevitable insurgent efforts to disrupt it. Any genuine grievances must be righted and injustices removed, so that every loyal citizen feels that those in power have his interests at heart. The Security Forces can contribute by showing good discipline, understanding and restraint in their dealings with the people. The natural humanity of the British Serviceman is justly famed worldwide and has achieved more goodwill than can be calculated.

Enlightened rule along these lines, particularly when it is by a national Government as opposed to an external controlling Power, is of great importance, for it will not only develop general confidence in the authorities, but will at the same time thwart the usual exploitation by the insurgents of any national grievances or weaknesses.

The third factor is the building up of public confidence by the establishment ultimately of a sound national Government, even if it is at first under the auspices of a controlling Power. Britain was able to do this so successfully in Malaya that it was possible to hand over control to the Malays, even before the Emergency was ended. The United States are trying to do the same in South

Vietnam, but face great difficulties due to the lack of any firm foundation on which to build.

The fourth and perhaps the best-known measure for winning hearts and minds is what the Americans call 'civic action'. This is organized aid to a country in a variety of forms, and is now an accepted feature of most counter-insurgency campaigns. It covers direct financial aid, technical advice and the provision of amenities such as schools, hospitals, roads, new industries, food and welfare. 'Operation Service' by the Police during the Malayan Emergency[1] was a good example of 'civic action', as were the policies of 'resettlement' and 'new villages'. Again, the Services can and do help considerably by providing medical and engineer assistance, and showing a real interest in the people and their problems.

A most valuable form of 'civic action' is the training of the populace to help themselves when they finally achieve their freedom from the insurgency. In a newly independent country, trained doctors and nurses may be worth more than understaffed hospitals; qualified administrators, a sound Civil Service, honest judges and trained schoolteachers are all essential to the country's future survival and well-being. It is not a quick process, but it produces valuable long-term benefits.

'Civic action' measures serve also to emphasize the contrast between constructive democracy and destructive Communism, and to expose the cynical mockery of most insurgent promises. They can do much to persuade the waverers that they would be better off supporting the Government rather than the opposition.

Finally, the Public Relations and Information aspects are all-important. Not only must the propaganda, indoctrination and falsehoods of the insurgents be countered quickly and firmly, but the Government must also put across their own case forcibly and frequently. The local population will be subjected to powerful enemy propaganda and pressure, and the authorities must deploy and use all possible means, in order to defeat this and at the same time to persuade the people that their best interests lie with the Government.

[1] See p. 67.

Propaganda is an unpopular word, but it is a powerful weapon for either side in any cold war campaign and any insurgency, and it is one that Britain could wield with more skill and conviction. It can be just as effective as armies in defeating insurgency—and very much cheaper!

CONCLUSION

The Communist bloc have long made a thorough study of insurgency as a highly effective weapon in the cold war, which they will doubtless continue to fight for some years yet. Rebel leaders worldwide have also studied this form of warfare to see how it can be used to achieve their ambitions, whatever they may be. It is virtually certain, therefore, that Britain and her Allies can expect to face further insurgencies in some form or another in the future.

This is now generally accepted, and greater emphasis is very rightly being given in the British Army to the requirements for counter-insurgency operations, with regard to organization, equipment and training. It is a subject which is constantly changing its form, and it calls for continuous and intensive study, if we are to be one move ahead of the opposition instead of dancing to their tune. It is hoped that the lessons of the past, set out in this book, may be of value in future counter-insurgency campaigning.

BRITISH ARMY OPERATIONS (1945–66) *Appendix A*

Year	*Limited War*	*Counter-Insurgency Campaigns*			*Policing Operations*				
1945		Greece							
1946				Palestine					
1947		Aden							
1948		Malaya							British Honduras
1949									
1950	Korea				Singapore				
1951					Akaba				
1952				Kenya					
1953					British Guiana				
1954			Cyprus						
1955					Buraimi				
1956	Suez				Hong Kong				
1957				Togoland					British Honduras
				Muscat & Oman					
1958					Aden			Jordan	Nassau
1959									
1960					Cameroons				Jamaica
1961	Kuwait								Zanzibar
1962		Brunei							British Guiana
								British Honduras	
1963		Malaysia		Aden	Cyprus	Swaziland	Zanzibar		
1964							East Africa		
1965							Mauritius		
1966									

INSURGENCIES WORLDWIDE (1945–66)[1]

Serial	Year	Country	Motive	Insurgent Forces	Insurgent Leader	Counter-Insurgent Forces	Did Insurgents win?	Insurgent Sanctuary
1	1927–50	China	Communism	Chinese Communist Army	Mao Tse-tung	Nationalists	Yes	
2	1944–45	Greece I	Communism	ELAS	Nikos Zachariades	Britain	No	
3	1946–49	Greece II	Communism	ELAS	?	Greece	No	Yugoslavia, Albania
4	1946–49	Palestine	Nationalism	IZL & Hagana		Britain	Yes	
5	1946–49	Philippines	Communism	Huks	Dr. J. Lava	Philippines	No	
6	1946–54	Indo-China	Communism	Viet Minh	Ho Chi Minh	France	Yes	China
7	1957–?	Vietnam	Communism	Viet Cong	Nguyen Huu Tho	U.S.A. and S. Vietnam	?	North Vietnam
8	1946–54	Laos I	Communism	Pathet Lao	Kayson Phomiuhane	France	No	
9	1960–?	Laos II	Communism	Pathet Lao		Laos	?	North Vietnam
10	1948–60	Malaya	Communism	MRLA	Chin Peng	Britain	No	
11	1952–56	Kenya	Tribal	Mau Mau	Dedan Kimathi	Britain	No	
12	1954–63	Algeria	Nationalism	ALN	Ferhet Abbas	France	Yes	Tunisia
13	1955–59	Cyprus	Nationalism	EOKA	Grivas	Britain	No	
14	1956	Oman	Tribal			Britain	No	
15	1960–?	Angola	Nationalism		Holden Roberto	Portugal		Congo
16	1962–63	Brunei	Nationalism			Britain	No	
17	1963–66	Malaysia	External aggression	Indonesian		Malaysia and Britain	No	Indonesia
18	1953–60	Cuba	Communism		Castro	Cuba	Yes	
19	1962–?	Yemen	Political	Royalist forces	Imam Badr	Egypt		
20	1963–?	South Arabia	Political	FLN		Britain		Yemen

[1] With acknowledgement to the *Army Quarterly*, January 1966.

Appendix C

OUTLINE INSURGENT ORGANIZATIONS

Basic Pattern		MALAYA		KENYA		CYPRUS	
Political	Military	Political	Military	Political	Military	Political	Military
(a)	(b)	(c)	(d)	(e)	(f)	(g)	(h)
Political Leader	*Military Leader*	Chin Peng	Lav Yew	?	Dedan Kimathi	Archbishop Makarios	General Grivas
Political Headquarters	*Military Headquarters*	Central Committee	HQ Malayan Races Liberation Army	Central Committee	HQ Land Freedom Army	Central Committee	Leader's HQ
Regional Organization	*Regional Headquarters*	State Committees	Regiments	District Committees	Armies	PEKA Committees	Sector HQ
		District Committees					
Local Branches	*Front Line Units*	Branches	Platoons	Passive Wing	Platoons	Branches	Groups

ORGANIZATIONS IN THE MALAYAN EMERGENCY *Appendix D*

Level	*Civil Administration*	*Police*	*Command and Control Structure (from 1950)*	*Malayan Communist Party*
(*a*)	(*b*)	(*c*)	(*d*)	(*e*)
Overall	High Commissioner	Commissioner of Police	High Commissioner's Daily Co-ordinating Committee	Central Committee
Federation	Executive Council Legislative Council	Federal Police Force	Emergency Operations Council (EOC) Director of Operations Committee (DOC)	Regional Bureaux (North Malaya and South Malaya)
State	State Council	Contingent (Chief Police Officer)	State War Executive Committee (SWEC)	State Committee
District	District Officer	Circles (Officers Superintending Police Circles)	District War Executive Committee (DWEC) Operations Committee	District Committee
Sub-District	Penghulu/Penggawa	Districts (Officers Commanding Police Districts)	Joint Operations Rooms	Branch Committee

Appendix E

COMMAND AND CONTROL STRUCTURE FOR THE CYPRUS EMERGENCY

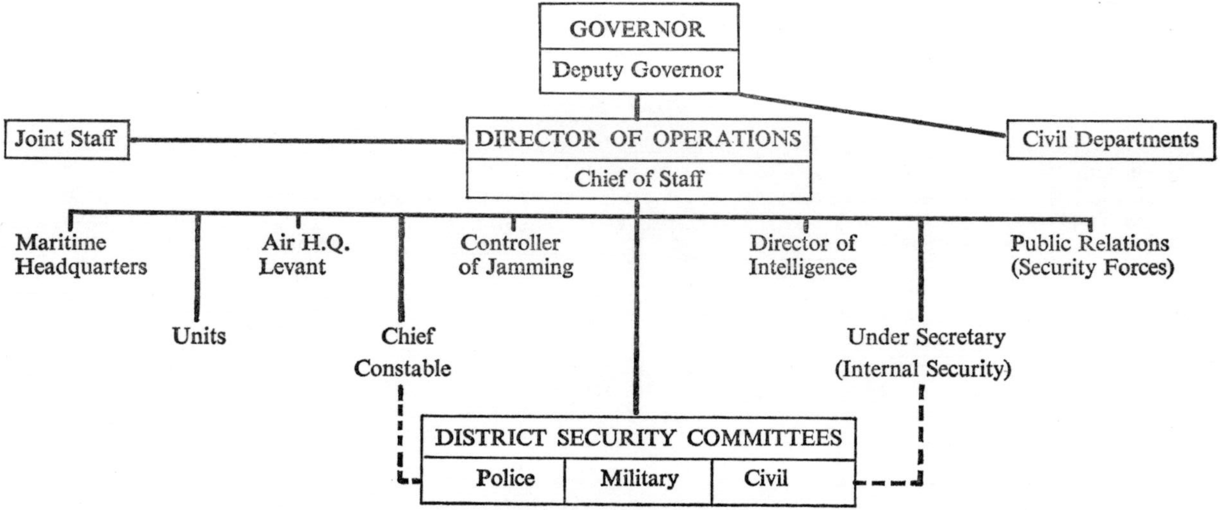

BIBLIOGRAPHY

This list of books does not cover all the documents read in the course of writing this book, but is intended to provide a guide to those who wish to make a further study of the subject. Articles in various civil and military magazines are not included.

GENERAL

Defeating Communist Insurgency	Sir Robert Thompson	1966
Partisan Warfare	Otto Heilbrunn	1962
The Cold War—A Re-Appraisal	E. Luard	1964
Guerrilla	C. W. Thayer	1964
Defence in the Cold War	Chatham House	1950
United Nations Forces	D. W. Bowett	1964
Peace-Keeping by the U.N. Forces	Burns and Heathcote	1963
Selected Works of Mao Tse-tung		1954

MALAYA

South East Asia in Turmoil	Brian Crozier	1966
Shoot to Kill	R. Miers	1959
Report from Malaya	Vernon Bartlett	1954
Templer in Malaya	C. N. Parkinson	1954
Red Shadow over Malaya	M. C. A. Henniker	1955

KENYA

Origins and Growth of Mau Mau (*The Corfield Report*)	HMSO Cmd 1030	1960

Report by Parliamentary Delegation to Kenya	HMSO Cmd 9081	1954
So Rough A Wind	Michael Blundell	1964
State of Emergency	F. Majdalany	1962
The Hunt for Kimathi	Henderson and Goodhart	1958
Gangs and Counter-Gangs	F. E. Kitson	1960
Defeating Mau Mau	L. S. B. Leakey	1954

CYPRUS

Guerrilla Warfare	General Grivas	1964
Island in Revolt	Charles Foley	1962
Terrorism in Cyprus—the Captured Documents	Colonial Office	1956
Cyprus Guerrilla	D. Alostos	1960
Grivas and the Story of EOKA	W. Byford-Jones	1959
Cyprus, the Dispute and Settlement	Chatham House	1959
Cyprus and Makarios	Stanley Mayes	1960
Bitter Lemons	L. Durrell	1957

INDEX

Index